California SAXON MATH™

Intermediate 4

Student Edition

Volume 2

T 330294

Stephen Hake

SAXON™

A Harcourt Achieve Imprint

www.SaxonPublishers.com
1-800-284-7019

ACKNOWLEDGEMENTS

This book was made possible by the significant contributions of many individuals and the dedicated efforts of talented teams at Harcourt Achieve.

Special thanks to Chris Braun for conscientious work on Power Up exercises, Problem Solving scripts, and student assessments. The long hours and technical assistance of John and James Hake were invaluable in meeting publishing deadlines. As always, the patience and support of Mary is most appreciated.

– Stephen Hake

Staff Credits

Editorial: Joel Riemer, Paula Zamarra, Hirva Raj, Smith Richardson, Pamela Cox, Michael Ota, Stephanie Rieper, Ann Sissac, Gayle Lowery, Robin Adams, David Baceski, Brooke Butner, Cecilia Colome, James Daniels, Leslie Bateman, Chad Barrett, Heather Jernt

Design: Alison Klassen, Joan Cunningham, Alan Klemp, Julie Hubbard, Lorelei Supapo, Andy Hendrix, Rhonda Holcomb

Production: Mychael Ferris-Pacheco, Jennifer Cohorn, Greg Gaspard, Donna Brawley, John-Paxton Gremillion

Manufacturing: Cathy Voltaggio, Kathleen Stewart

Marketing: Marilyn Trow, Kimberly Sadler

E-Learning: Layne Hedrick

ABOUT THE AUTHOR

Stephen Hake has authored six books in the **Saxon Math** series. He writes from 17 years of classroom experience as a teacher in grades 5 through 12 and as a math specialist in El Monte, California. As a math coach, his students won honors and recognition in local, regional, and statewide competitions.

Stephen has been writing math curriculum since 1975 and for Saxon since 1985. He has also authored several math contests including Los Angeles County's first Math Field Day contest. Stephen contributed to the 1999 National Academy of Science publication on the Nature and Teaching of Algebra in the Middle Grades.

Stephen is a member of the National Council of Teachers of Mathematics and the California Mathematics Council. He earned his BA from United States International University and his MA from Chapman College.

CONTENTS OVERVIEW

Integrated and Distributed Units of Instruction

California Strands Key:
NS = Number Sense
AF = Algebra and Functions
MG = Measurement and Geometry

SDAP = Statistics, Data Analysis, and Probability
MR = Mathematical Reasoning

TABLE OF CONTENTS

Section 3 — Lessons 21–30, Investigation 3

California Strands Key:
NS = Number Sense
AF = Algebra and Functions
MG = Measurement and Geometry
SDAP = Statistics, Data Analysis, and Probability
MR = Mathematical Reasoning

TABLE OF CONTENTS

Section 5 *Lessons 41–50, Investigation 5*

California Strands Key:
NS = Number Sense
AF = Algebra and Functions
MG = Measurement and Geometry

SDAP = Statistics, Data Analysis, and Probability
MR = Mathematical Reasoning

TABLE OF CONTENTS

Section 7 *Lessons 61–70, Investigation 7*

California Strands Key:
NS = Number Sense
AF = Algebra and Functions
MG = Measurement and Geometry
SDAP = Statistics, Data Analysis, and Probability
MR = Mathematical Reasoning

Section 9 — *Lessons 81–90, Investigation 9*

California Strands Key:
NS = Number Sense
AF = Algebra and Functions
MG = Measurement and Geometry
SDAP = Statistics, Data Analysis, and Probability
MR = Mathematical Reasoning

TABLE OF CONTENTS

Section 11 *Lessons 101–110, Investigation 11*

California Strands Key:
NS = Number Sense
AF = Algebra and Functions
MG = Measurement and Geometry

SDAP = Statistics, Data Analysis, and Probability
MR = Mathematical Reasoning

TABLE OF CONTENTS

Dear Student,

We study mathematics because it plays a very important role in our lives. Our school schedule, our trip to the store, the preparation of our meals, and many of the games we play involve mathematics. The word problems in this book are often drawn from everyday experiences.

When you become an adult, mathematics will become even more important. In fact, your future may depend on the mathematics you are learning now. This book will help you to learn mathematics and to learn it well. As you complete each lesson, you will see that similar problems are presented again and again. *Solving each problem day after day is the secret to success.*

Your book includes daily lessons and investigations. Each lesson has three parts.

1. The first part is a Power Up that includes practice of basic facts and mental math. These exercises improve your speed, accuracy, and ability to do math *in your head.* The Power Up also includes a problem-solving exercise to help you learn the strategies for solving complicated problems.

2. The second part of the lesson is the New Concept. This section introduces a new mathematical concept and presents examples that use the concept. The Lesson Practice provides a chance for you to solve problems using the new concept. The problems are lettered a, b, c, and so on.

3. The final part of the lesson is the Written Practice. This section reviews previously taught concepts and prepares you for concepts that will be taught in later lessons. Solving these problems will help you practice your skills and remember concepts you have learned.

Investigations are variations of the daily lesson. The investigations in this book often involve activities that fill an entire class period. Investigations contain their own set of questions but do not include Lesson Practice or Written Practice.

Remember to solve every problem in each Lesson Practice, Written Practice, and Investigation. Do your best work, and you will experience success and true learning that will stay with you and serve you well in the future.

Temple City, California

HOW TO USE YOUR TEXTBOOK

Saxon Math Intermediate 4 is unlike any math book you have used! It doesn't have colorful photos to distract you from learning. The Saxon approach lets you see the beauty and structure within math itself. You will understand more mathematics, become more confident in doing math, and will be well prepared when you take high school math classes.

Power Yourself Up

Start off each lesson by practicing your basic skills and concepts, mental math, and problem solving. Make your math brain stronger by exercising it every day. Soon you'll know these facts by memory!

Learn Something New!

Each day brings you a new concept, but you'll only have to learn a small part of it now. You'll be building on this concept throughout the year so that you understand and remember it by test time.

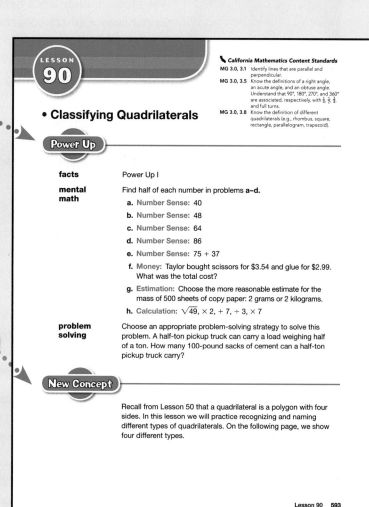

LESSON 90

California Mathematics Content Standards

MG 3.0, 3.1 Identify lines that are parallel and perpendicular.
MG 3.0, 3.5 Know the definitions of a right angle, an acute angle, and an obtuse angle. Understand that 90°, 180°, 270°, and 360° are associated, respectively, with $\frac{1}{4}$, $\frac{1}{2}$, $\frac{3}{4}$, and full turns.
MG 3.0, 3.8 Know the definition of different quadrilaterals (e.g., rhombus, square, rectangle, parallelogram, trapezoid).

• Classifying Quadrilaterals

Power Up

facts   Power Up I

mental math   Find half of each number in problems **a–d.**

   a. Number Sense: 40

   b. Number Sense: 48

   c. Number Sense: 64

   d. Number Sense: 86

   e. Number Sense: 75 + 37

   f. Money: Taylor bought scissors for $3.54 and glue for $2.99. What was the total cost?

   g. Estimation: Choose the more reasonable estimate for the mass of 500 sheets of copy paper: 2 grams or 2 kilograms.

   h. Calculation: $\sqrt{49}$, × 2, + 7, ÷ 3, × 7

problem solving   Choose an appropriate problem-solving strategy to solve this problem. A half-ton pickup truck can carry a load weighing half of a ton. How many 100-pound sacks of cement can a half-ton pickup truck carry?

New Concept

Recall from Lesson 50 that a quadrilateral is a polygon with four sides. In this lesson we will practice recognizing and naming different types of quadrilaterals. On the following page, we show four different types.

Lesson 90   593

Activity 1

Quadrilaterals in the Classroom

Look around the room for quadrilaterals. Find examples of at least three different types of quadrilaterals illustrated in the beginning of this lesson. Draw each example you find, and next to each picture, name the object you drew and its shape. Then describe how you know that the object is the shape you named and describe the relationships of the sides of each quadrilateral.

Activity 2

Symmetry and Quadrilaterals

Materials needed:
- **Lesson Activity 25**
- mirror or reflective surface

If a figure can be divided into mirror images by a line of symmetry, then the figure has reflective symmetry. A mirror can help us decide if a figure has reflective symmetry. If we place a mirror upright along a line of symmetry, the half of the figure behind the mirror appears in the reflection of the other half. Use a mirror to discover which figures in **Lesson Activity 25** have reflective symmetry. If you find a figure with reflective symmetry, draw its line (or lines) of symmetry.

Lesson Practice

(Classify) Describe each quadrilateral as a trapezoid, parallelogram, rhombus, rectangle, or square. (More than one description may apply to each figure.)

a.   b.   c.   d.

e. Describe the angles in figures **a–d** and the relationships between the sides.

f. Draw two parallel line segments that are the same length. Then make a quadrilateral by drawing two more parallel line segments that connect the endpoints. Is your quadrilateral a parallelogram? Why or why not?

Get Active!

Dig into math with a hands-on activity. Explore a math concept with your friends as you work together and use manipulatives to see new connections in mathematics.

Check It Out!

The Lesson Practice lets you check to see if you understand today's new concept.

Written Practice   *Distributed and Integrated*

* 1. (Analyze) What is the total number of days in the first three months of a leap year?
(RF12)

* 2. Thirty-two desks were arranged as equally as possible in 6 rows.
(87)
 a. How many rows had exactly 5 desks?

 b. How many rows had 6 desks?

* 3. (Evaluate) If $y = 3x + 6$, then what is y when $x = 4$, 5, and 6?
(64,
Inv. 7)

* 4. (Analyze) Carmen separated the 37 math books as equally as possible into 4 stacks
(87)
 a. How many stacks had exactly 9 books?

 b. How many stacks had 10 books?

* 5. (Conclude) Write *true* or *false* for parts **a–e.**
(90)
 a. All rectangles have four right angles.

 b. Some squares are rectangles.

 c. All trapezoids are rhombuses.

 d. All rectangles are parallelograms.

 e. Some parallelograms have no right angles.

6. a. What decimal number names the shaded part of the large square at right?
(Inv. 4)

 b. What decimal number names the part that is not shaded?

* 7. (Explain) Near closing time, 31 children and adults are waiting in line to board a ride at an amusement park. Eight people board the ride at one time. How many people will be on the last ride of the day? Explain your answer.
(87)

Exercise Your Mind!

When you work the Written Practice exercises, you will review both today's new concept and also math you learned in earlier lessons. Each exercise will be on a different concept — you never know what you're going to get! It's like a mystery game — unpredictable and challenging.

As you review concepts from earlier in the book, you'll be asked to use higher-order thinking skills to show what you know and why the math works.

HOW TO USE YOUR TEXTBOOK

Become an Investigator!

Dive into math concepts and explore the depths of math connections in the Investigations.

Continue to develop your mathematical thinking through applications, activities, and extensions.

Focus on

Investigating Equivalent Fractions with Manipulatives

Fraction manipulatives can help us better understand fractions. In this investigation we will make and use a set of fraction manipulatives.

Activity 1

Using Fraction Manipulatives

Materials needed:
- **Lesson Activities 11, 12,** and **13**
- scissors
- envelopes or locking plastic bags (optional)

Model Use your fraction manipulatives to complete the following exercises:

1. Another name for $\frac{1}{4}$ is a quarter. How many quarters of a circle does it take to form a whole circle? Show your work.

2. Fit two quarter circles together to form a half circle. That is, show that $\frac{2}{4}$ equals $\frac{1}{2}$.

3. How many fourths equals $1\frac{1}{4}$?

4. This number sentence shows how to make a whole circle using half circles:

$$\frac{1}{2} + \frac{1}{2} = 1$$

Write a number sentence that shows how to make a whole circle using only quarter circles.

5. How many half circles equals $1\frac{1}{2}$ circles?

6. Four half circles make how many whole circles?

Model Manipulatives can help us compare and order fractions. Use your fraction manipulatives to illustrate and answer each problem:

7. Arrange $\frac{1}{2}$, $\frac{1}{8}$, and $\frac{1}{4}$ in order from least to greatest.

California Mathematics Content Standards

NS 1.0, 1.6 Write tenths and hundredths in decimal and fraction notations and know the fraction and decimal equivalents for halves and fourths (e.g., $\frac{1}{2} = 0.5$ or .50; $\frac{7}{4} = 1\frac{3}{4} = 1.75$).

NS 1.0, 1.7 Write the fraction represented by a drawing of parts of a figure; represent a given fraction by using drawings; and relate a fraction to a simple decimal on a number line.

MR 2.0, 2.3 Use a variety of methods, such as words, numbers, symbols, charts, graphs, tables, diagrams, and models, to explain mathematical reasoning.

Investigation 8 **537**

LESSON
61

\ *California Mathematics Content Standards*

NS 1.0, 1.3 Round whole numbers through the millions to the nearest ten, hundred, thousand, ten thousand, or hundred thousand.

NS 1.0, 1.4 Decide when a rounded solution is called for and explain why such a solution may be appropriate.

NS 2.0, 2.1 Estimate and compute the sum or difference of whole numbers and positive decimals to two places.

MR 2.0, 2.5 Indicate the relative advantages of exact and approximate solutions to problems and give answers to a specified degree of accuracy.

• Estimating Arithmetic Answers

Power Up

facts	Power Up I
count aloud	Count by halves from $\frac{1}{2}$ to 6 and back down to $\frac{1}{2}$.
mental math	Multiply a number by 10 in problems **a–c**.

  a. Number Sense: 12×10

  b. Number Sense: 120×10

  c. Number Sense: 10×10

  d. Money: Jill paid for a pencil that cost 36¢ with a $1 bill. How much change should she receive?

  e. Money: One container of motor oil costs $3.75. How much do 2 containers cost?

  f. Fractional Part: The whole circle has been divided into quarters. What fraction of the circle is shaded? What fraction is not shaded?

  g. Estimation: Phil plans to buy lasagna for $5.29 and a drink for $1.79. Round each price to the nearest 25 cents and then add to estimate the total cost.

  h. Calculation: $48 + 250 + 6 + 6$

problem solving

Choose an appropriate problem-solving strategy to solve this problem. Garcia is packing his clothes for summer camp. He wants to take three pairs of shorts. He has four different pairs of shorts from which to choose—tan, blue, white, and black. What are the different combinations of three pairs of shorts that Garcia can pack?

We can estimate arithmetic answers by rounding numbers. Estimating does not give us the exact answer, but it can give us an answer that is close to the exact answer. For some problems, an estimate is all that is necessary to solve the problem. When an exact answer is needed, estimating is a way to decide whether our exact answer is reasonable. Estimating is useful for many purposes, such as mentally adding price totals when shopping.

Example 1

Thinking Skills

Discuss

Which place is used to round a 3-digit number to the nearest hundred?

Estimate the sum of 396 and 512.

To estimate, we first round the number to the nearest hundred. We round 396 to 400 and 512 to 500. Then we find the estimated sum by adding 400 and 500.

$$
\begin{array}{r}
400 \\
+\ 500 \\
\hline
900
\end{array}
$$

The estimated sum of 396 and 512 is **900.** The exact sum of 396 and 512 is 908. The estimated answer is not equal to the exact answer, but it is close.

Example 2

Thinking Skills

Connect

Which place is used to round a 2-digit number to the nearest ten?

Estimate the product of 72 and 5.

We round the two-digit number, but we generally do not round a one-digit number when estimating. The estimated product of 72 and 5 is **350.**

$$
\begin{array}{r}
70 \\
\times\ \ 5 \\
\hline
350
\end{array}
$$

The exact product of 72 and 5 is 360. The estimated product is a little less than the exact answer, 360, because 72 was rounded down to 70 for the estimate.

Example 3

To estimate 7 × 365, Towanda multiplied 7 by 400. Was Towanda's estimate more than, equal to, or less than the actual product of 7 and 365?

Towanda's estimate was **more than the actual product** of 7 and 365 because she rounded 365 up to 400 before multiplying.

Example 4

Estimate the answer to 43 ÷ 8.

To estimate division answers, we want to use numbers that divide easily, so we change the problem slightly. We keep the number we are dividing by, which is 8, and we change the number that is being divided, which is 43, to a compatible number. We change 43 to a nearby number that can be divided easily by 8, such as 40 or 48. Using 40, the estimated answer is **5.** Using 48, the estimated answer is **6.** Since 43 is between 40 and 48, the actual answer is more than 5 but less than 6. That is, the exact answer is 5 plus a remainder.

Example 5

Nicola wants to buy a box of cereal for $5.89, a gallon of milk for $3.80, and a half gallon of juice for $2.20. Nicola has $13.00. Does she have enough to pay for the groceries?

Since we don't need an exact answer we can round the prices of each item to the nearest dollar.

The cereal cost $5.89, which is closer to $6 than to $5.

The milk cost $3.80, which is closer to $4 than to $3.

The juice cost $2.20, which is closer to $2 than to $3.

Thinking Skills

Verify

How do we round $3.80 to the nearest dollar? Explain your thinking.

Item	Price	Rounded to the Nearest Dollar
cereal	$5.89	$6
milk	$3.80	$4
juice	$2.20	$2

To estimate the total, we add the rounded numbers.

$$\$6 + \$4 + \$2 = \$12$$

Nicola's estimated grocery bill was **about $12.**

Explain Suppose that Nicola wanted to be sure she had enough money to purchase all of the items *before* she reached the checkout line. How should she round the prices? Explain your reasoning.

Lesson Practice Estimate the answer to each arithmetic problem. Then find the exact answer.

a. 59 + 68 + 81   **b.** 607 + 891

c. 585 − 294   **d.** 82 − 39

e. 59 × 6   **f.** 397 × 4

g. 42 ÷ 5   **h.** 29 ÷ 7

i. (Explain) Dixie estimated the product of 5 and 5280 by multiplying 5 by 5000. Was Dixie's estimate more than, equal to, or less than the actual product? Why?

j. Mariano would like to purchase a notebook computer, a wireless mouse, and an accessory carrying bag. The cost of each item is shown in the table.

Item	Cost
Notebook computer	$845
Wireless mouse	$27.50
Accessory bag	$39.95

What is a reasonable estimate of Mariano's total cost? Explain your thinking.

Written Practice

Distributed and Integrated

*1.
(58)
A comfortable walking pace is about 3 miles per hour. How far would a person walk in 4 hours at a pace of 3 miles per hour? Make a table to solve the problem.

2.
(60)
The Johnson Family drank 33 glasses of milk in 3 days. How many glasses of milk is that each day?

*3. a. **Analyze** Find the perimeter and area of this rectangle. Remember to
(Inv. 3, 31) label your answer with "units" or "square units".

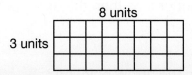

8 units

3 units

b. **Represent** Sketch a rectangle that is four units wide
with the same area as the rectangle in part **a.** What is the
perimeter of this new rectangle?

4. **Multiple Choice** Which of these numbers is *not* a factor of 12?
(55)
A 6 **B** 5 **C** 4 **D** 3

5. The starting time was before sunrise. The stopping time was in the
(13) afternoon. What was the difference between the two times?

Starting time   Stopping time

6. **Represent** One square mile is 3,097,600 square yards. Use words
(47) to write that number of square yards.

7. a. What fraction of this pentagon is not shaded?
(57)

b. Is the shaded part of this pentagon more than $\frac{1}{2}$ or
less than $\frac{1}{2}$ of the pentagon?

8. According to this calendar, what is the date of the last
(RF12) Saturday in July 2019?

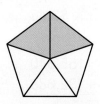

JULY 2019						
S	M	T	W	T	F	S
	1	2	3	4	5	6
7	8	9	10	11	12	13
14	15	16	17	18	19	20
21	22	23	24	25	26	27
28	29	30	31			

* 9. **Estimate** To estimate the product of two factors, a student
(61) rounded one factor down and left the other factor unchanged. Was the
estimate greater than the exact product or less than the exact product?
Give an example to support your answer.

10. **Represent** To what mixed number is the arrow pointing?
(32)

7   8

* 11. **Justify** Sofia estimated that the exact product of 4 × 68 is close to
(61) 400 because 68 rounded to the nearest hundred is 100, and 4 × 100 = 400.
Was Sofia's estimate reasonable? Explain why or why not.

* 12. **Represent** Which of the following is the number 3,003,016?
(47)
    A three million, three hundred sixteen
    B three million, three thousand, sixteen
    C three hundred million, sixteen
    D three hundred million, three thousand, sixteen

13. $6.25 + $4 + $12.78
(28)

14. 3.6 + 12.4 + 0.84
(45)

15. $30.25
(8, 28, 52) −   _b_
       $13.06

16.   149,384
(52) − 98,765

* 17.   409
(59) ×   7

18. 5 × $3.46
(59)

19. $0.79 × 6
(59)

20.   155,340
(51) + 32,688

21.   600
(37) ×   6

* 22.   607
(59) ×   3

* 23. 45 ÷ 6
(54)

* 24. $\frac{83}{9}$
(54)

* 25. 7)‾60
(54)

*26. This line plot shows the number of times some students visit the city
(Inv. 6) zoo each year. Use this line plot to answer parts **a–c.**

Number of Visits to the City Zoo

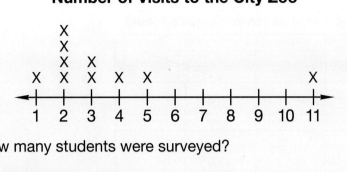

  a. How many students were surveyed?

  b. What is the mode?

  c. Is there an outlier? If yes, what is it?

27. Name the shaded part of this rectangle as a fraction and as a decimal.
(Inv. 4)

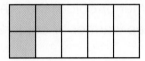

28. How much money is $\frac{1}{4}$ of a dollar?
(29)

*29. (**Represent**) Draw a hexagon. A hexagon has how many vertices?
(50)

*30. **Interpret** The line graph shows the temperature at different times
(Inv. 5)   on a winter morning at Hayden's school. Use the graph to answer the
questions that follow.

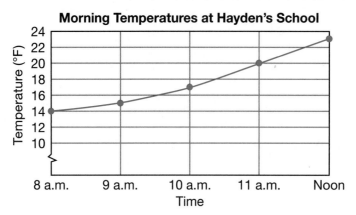

a. At what time was the first temperature of the morning recorded?
What was that temperature?

b. Was the noon temperature warmer or colder than the 10 a.m.
temperature? How many degrees warmer or colder?

*Real-World
Connection*

How many line segments make up a pentagon? That's easy,
right? The answer is 5. How many line segments make
up a pentacontagon? Not so easy. The answer is 50. A
pentacontagon is not commonly known, but it is indeed a
polygon. Refer to the table to learn the names and attributes of
less commonly known polygons.

Name of Polygon	Number of Sides
heptagon	7
nonagon	9
hendecagon	11
tridecagon	13
pentadecagon	15

a. Name a polygon that has 8 less sides than a tridecagon.

b. Name a polygon that has more sides than a heptagon but
fewer sides than a nonagon.

c. Name a polygon that has 12 more sides than a triangle.

d. Choose two polygons from the table above and draw models
of them.

✎ *California Mathematics Content Standards*

NS 1.0, ❶.❾ Identify on a number line the relative position of positive fractions, positive mixed numbers, and positive decimals to two decimal places.

NS 2.0, 2.1 Estimate and compute the sum or difference of whole numbers and positive decimals to two places.

NS 2.0, 2.2 Round two-place decimals to one decimal or the nearest whole number and judge the reasonableness of the rounded answer.

MR 2.0, 2.1 Use estimation to verify the reasonableness of calculated results.

• Rounding to the Nearest Tenth

facts	Power Up I
count aloud	Count by fourths from $5\frac{1}{4}$ to 10.
mental math	

a. Number Sense: 14×10

b. Money: Sean bought a ream of paper for $6.47 and a box of staples for $1.85. What was the total cost?

c. Fractional Parts: Compare: $\frac{1}{4}$ ◯ $\frac{1}{2}$

d. Geometry: What is the perimeter of a square that is 6 inches on each side?

e. Time: Crystal phoned her friend at 4:05 p.m. They talked for 22 minutes. What time did Crystal's phone call end?

f. Measurement: Ray cut a 1-foot length of string from a larger piece that was 22 inches long. How many inches of string remained?

g. Estimation: Washington School has 258 students. Lincoln School has 241 students. Round each number to the nearest ten and then add to estimate the total number of students.

h. Calculation: $400 + 37 + 210 - 17$

problem solving

Choose an appropriate problem-solving strategy to solve this problem. The following page shows a sequence of triangular numbers. The third term in the sequence, 6, is the number of dots in a triangular arrangement of dots with three rows. Notice that in this sequence the count from one number to the next increases. Find the number of dots in a triangular arrangement with 8 rows. Explain how you arrived at your answer and how you can verify your answer.

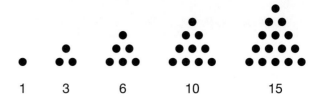

1   3   6   10   15

New Concept

We learned to use a number line to round a two place decimal to the nearest whole number. We can also use a number line to round a two place decimal to the nearest tenth. This number line is divided into ten equal segments. In Lesson 44 we learned that 1.2 and 1.20 are **equivalent decimals,** which is shown in the art below.

The decimal number 1.26 is between 1.2 and 1.3. We can see that 1.26 is closer to 1.3 than it is to 1.2. So, 1.26 rounded to the nearest tenth is 1.3.

Explain What is 1.25 rounded to the nearest tenth? Explain why.

Example 1

Round 7.52 to the nearest tenth.

Rounding 7.52 to the nearest tenth is the same as rounding $7.52 to the nearest ten cents. Just as $7.52 is between $7.50 and $7.60, 7.52 is between 7.5 and 7.6, as shown on the number line below.

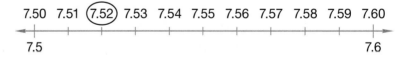

We can see that 7.52 is closer to 7.5 than 7.6, so, 7.52 rounded to the nearest tenth is **7.50 or 7.5.**

Example 2

Last month, Mr. Garcia paid $40.68 for a tank of gasoline. This month he paid $42.43 for a tank of gas. To the nearest ten cents, how much more did he pay this month?

Since we are asked to find the amount to the nearest ten cents, we will round both numbers to the nearest dime. Then we can subtract.

$42.43 rounds to $42.40

$40.68 rounds to $40.70

$$\begin{array}{r} \$42.40 \\ -\$40.70 \\ \hline \$1.70 \end{array}$$

To the nearest ten cents Mr. Garcia paid **$1.70** more for a tank of gas this month.

Analyze How much more did Mr. Garcia pay to the nearest quarter?

Example 3

Mrs. Jensen bought 1.33 pounds of cheddar cheese and 1.86 pounds of Swiss cheese. To the nearest tenth of a pound, how much cheese did she buy altogether?

Since we are asked to find the amount to the nearest tenth of a pound, we need to round both numbers to the nearest tenth. Then we can add.

1.33 rounds to 1.30

1.86 rounds to 1.90

$$\begin{array}{r} 1.30 \\ +1.90 \\ \hline 3.20 \end{array}$$

To the nearest tenth of a pound, Mrs. Jensen bought **3.2 pounds** of cheese.

Analyze How can you check that the estimate is reasonable?

Lesson Practice Round each number to the nearest ten cents:

a. $4.79   **b.** $3.25   **c.** $0.33

Round each number to the nearest tenth:

d. 5.43   **e.** 0.47   **f.** 3.62

Round each number to the nearest quarter:

g. $0.89   **h.** $1.46   **i.** $1.10

j. Beth and Tamarra competed in a 100-meter freestyle swim race. Beth finished in 51.9 seconds. Tamarra finished in 50.39 seconds. To the nearest tenth of a second, how much faster was Tamarra?

*1. (58) **Analyze** Alphonso ran 6 miles per hour. At that rate, how far could he run in 3 hours? Make a table to solve this problem.

2. (20) Find the perimeter and area of this rectangle:

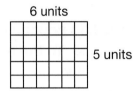

6 units

5 units

3. (Inv. 4) **Represent** Aletta ran 100 meters in twelve and fourteen hundredths seconds. Use digits to write her time.

4. (50) Taydren drew an octagon and a pentagon. How many sides did the two polygons have altogether?

*5. (59) 470×3 ***6.** (59) 6×394 ***7.** (59) 856×4

*8. (Inv. 6) **Interpret** Use this set of data to answer parts **a** and **b**.

5, 3, 6, 5, 9, 1, 5, 4, 3

a. What is the median?

b. What is the mode?

9. (32) **Represent** To what mixed number is the arrow pointing?

6   7   8

*10. (32) **Model** Draw a number line and show the locations of 0, 1, 2, $1\frac{2}{3}$, and $2\frac{1}{3}$.

11. (27) **Represent** Mount Rainier stands four thousand, three hundred ninety-two meters above sea level. Use digits to write that number.

12. Mo'Nique could make 35 knots in 7 minutes. How many knots could she make in 1 minute?
(60)

13. Estimate the sum of 6810 and 9030 by rounding each number to the nearest thousand before adding.
(61)

*** 14.** Estimate the sum of $12.15 and $5.95. Then find the exact sum.
(61)

15. $20 − ($8.95 + 75¢)
(9, 28)

16. 23.64 − 5.45
(45)

17.   43¢
(38)   × 8

18. $3.05
(59)   × 5

19. $2.63
(59)   × 7

20. (Connect) Rewrite this addition problem as a multiplication problem and find the answer:
(23)

$$64 + 64 + 64 + 64 + 64$$

*** 21.** 5)47
(54)

*** 22.** 7)65
(54)

*** 23.** 3)26
(54)

*** 24.** $\frac{39}{6}$
(54)

*** 25.** 4r = 48
(54)

*** 26.** 46 ÷ 8
(54)

27. (Model) Use an inch ruler to find the lengths of segments AB, BC, and AC.
(33, 42)

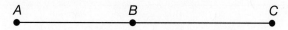

*** 28.** **Multiple Choice** Which of the following is 6.25 rounded to the nearest tenth?
(62)

    A 6.0   **B** 6.10   **C** 6.20   **D** 6.30

*** 29.** (Represent) How many yards are equal to 5280 ft?
(42)

*30. **Interpret** The lengths of three land tunnels in the United States are
(27, 42, Inv. 5) shown in the graph. Use the graph to answer parts **a–c.**

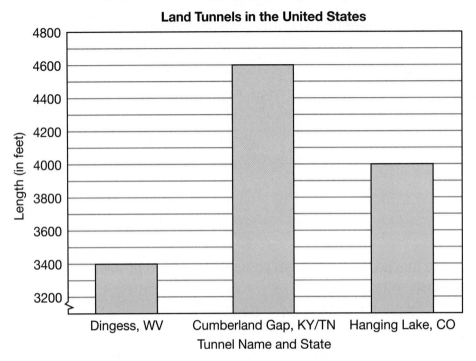

Land Tunnels in the United States

a. Write the names of the tunnels in order from shortest to longest.

b. How many feet longer is the Hanging Lake Tunnel than the Dingess Tunnel?

c. Is the Cumberland Gap tunnel greater or less than one mile long?

• Order of Operations, Part 3

California Mathematics Content Standards

AF 1.0, **1.2** Interpret and evaluate mathematical expressions that now use parentheses.

AF 1.0, **1.3** Use parentheses to indicate which operation to perform first when writing expressions containing more than two terms and different operations.

AF **2.0, 2.1** Know and understand that equals added to equals are equal.

AF **2.0, 2.2** Know and understand that equals multiplied by equals are equal.

facts	Power Up J
count aloud	Count down by thousands from 20,000 to 1000.
mental math	Multiply three numbers in problems **a–c.**

a. Number Sense: $6 \times 7 \times 10$

b. Number Sense: $5 \times 8 \times 10$

c. Number Sense: $12 \times 10 \times 10$

d. Money: $\$7.59 + \$0.95$

e. Money: Sydney had $5.00. Then she spent $3.25 on photocopies. How much money does she have left?

f. Geometry: Compare: $4\frac{1}{2}$ in. $\bigcirc$ radius of a circle with a 10 in. diameter

g. Estimation: Henry estimated that his full drinking glass contained 400 mL of water. Is this a reasonable estimate?

h. Calculation: $470 - 30 + 62 + 29$

problem solving	Choose an appropriate problem-solving strategy to solve this problem. Half of the students in Gabriel's class are girls. Do we know how many students are in this class? Do we know whether there are more boys or more girls in the class? Do we know whether the number of students in the class is even or odd?

We have learned some rules about the order of operations.

Order of Operations

- Complete operations inside parentheses first.
- Multiply and divide from left to right.
- Add and subtract from left to right.

We follow the order of operations whenever we simplify expressions or solve equations. Notice that the order of operations indicates that we multiply and divide from left to right before we add and subtract. Today, we are going to simplify expressions using all four operations.

When an expression includes both division and subtraction, we divide before we subtract.

In this expression, first we divide 24 by 4. Then we subtract 2.

$$24 \div 4 - 2$$

Step 1   Step 2

Step 1: $24 \div 4 = 6$
Step 2: $6 - 2 = \mathbf{4}$

If we want the steps done in a different order, we use parentheses to show which operation should be completed first.

In the expression below, the parentheses tell us to subtract first and then divide.

$$24 \div (4 - 2)$$

Step 2   Step 1

Step 1: $4 - 2 = 2$
Step 2: $24 \div 2 = \mathbf{12}$

Example 1

Simplify each expression:

a. $35 + 7 \times 3$ b. $56 \div (4 + 3)$ c. $35 - 5 \times 2$

We can use the order of operations to simplify each expression.

a. We multiply before we add.

$35 + 7 \times 3$ First multiply 7 by 3.

$35 + 21$ Then add 35 and 21.

56   The answer is 56.

b. We complete the operation inside the parentheses first.

$$56 \div (4 + 3) \qquad \text{First add 4 and 3.}$$
$$56 \div 7 \qquad \text{Then divide 56 by 7.}$$
$$\mathbf{8} \qquad \text{The answer is 8.}$$

c. We multiply before we subtract.

$$\mathbf{35 - 5 \times 2} \qquad \text{First multiply 5 by 2.}$$
$$35 - 10 \qquad \text{Then subtract 10 from 35.}$$
$$\mathbf{25} \qquad \text{The answer is 25.}$$

(**Discuss**) Where would we place parentheses in the expression $35 - 5 \times 2$ so it will simplify to an answer of 60?

Example 2

Evaluate $(20 \div 5) + (2 \times 3) - m$ when $m = 7$.

$$(20 \div 5) + (2 \times 3) - m \qquad \text{Substitute 7 for } m.$$
$$(20 \div 5) + (2 \times 3) - 7 \qquad \text{Divide 20 by 5.}$$
$$4 + (2 \times 3) - 7 \qquad \text{Multiply 2 by 3.}$$
$$4 + 6 - 7 \qquad \text{Add and subtract from left to right.}$$
$$\mathbf{3} \qquad \text{The answer is 3.}$$

Example 3

Solve each equation.

    a. If $5 \times 2 + 10 = 100 \div 10 + r$, what does r equal?

    b. If $(15 - 8) \times 9 = (49 \div 7) \times d$, what does d equal?

    a. The equals sign shows us that the quantities on both sides of the equation are equal.

$$5 \times 2 + 10 = 100 \div 10 + r \qquad \text{Multiply and divide from left to right.}$$
$$10 + 10 = 10 + r \qquad \text{For both sides of the equation to be equal, } r \text{ must equal 10.}$$
$$10 + 10 = 10 + 10 \qquad \text{Substitute 10 for } r.$$
$$20 = 20 \qquad \text{Check.}$$

We can see that $r = \mathbf{10}$ because we added the same number to equal amounts.

b. $(15 - 8) \times 9 = (49 \div 7) \times d$ Simplify inside the parentheses first.

$$7 \times 9 = 7 \times d$$ For both sides of the equation to be equal, d must equal 9.

$$7 \times 9 = 7 \times 9$$ Substitute 9 for d.

$$63 = 63$$ Check.

We can see that **$d = 9$** because we multiplied equal amounts by the same number.

Lesson Practice

Connect Simplify each expression. Remember to use the order of operations.

a. $4 \times 2 + 16 \div 8$ **b.** $27 \div 3 - 2 + 7$

c. $6 + (10 - 4) \times 3$ **d.** $9 \times (12 - 3) - 5$

Evaluate Simplify each expression when $k = 5$.

e. $20 - 10 \div k$ **f.** $6 \times (2 + k) - 5$

g. **Explain** If $s + 50 = t + 50$, is the equation below true? Why or why not?

$$s \times 10 = t \times 10$$

Written Practice   *Distributed and Integrated*

Formulate Write and solve equations for problems **1** and **2**.

1. Celeste has three hundred eighty-four baseball cards. Nathan has two
(15) hundred sixty baseball cards. Celeste has how many more cards than Nathan?

2. Forty-two students could ride in one bus. There were 3 buses. How
(39) many students could ride in all the buses?

*3. Jazmyn's house key is 4 cm long. How many millimeters long is her
(42) house key?

*4. **Represent** Write a decimal and a fraction (or a mixed number) to
(43) represent each point.

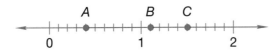

5. **Represent** Copy this hexagon and shade one sixth of it.
(19)

*6. a. This toothpick is how many centimeters long?
(42)

   b. This toothpick is how many millimeters long?

*7. **Interpret** Use this set of data to answer parts **a–c.**
(Inv. 6)

16, 32, 24, 60, 24, 16, 31, 19, 20

   a. What is the mode?

   b. What is the median?

   c. Is there an outlier? Name it.

8. **Analyze** If each side of a square is 1 yard long, then what is the
(20, 42) perimeter of the square in feet?

*9. **Explain** The number of students enrolled at each of three elementary
(61) schools is shown in the table below.

Elementary School Enrollment

School	Number of Students
Van Buren	412
Carter	495
Eisenhower	379

Use rounding to make a reasonable estimate of the total number of
students enrolled at the three schools. Explain your answer.

*10. Segment *AB* is 3.5 cm long. Segment *AC* is 11.6 cm long. How long
(33, 45) is segment *BC*? Write a decimal subtraction equation and find the
answer.

11. a. Hugo rode 60 miles in 5 hours. His average speed was how many miles per hour?
(58, 60)

    b. Hugo could ride 21 miles in 1 hour. At that rate, how many miles could Hugo ride in 7 hours?

* 12. The first three prime numbers are 2, 3, and 5. What are the next three prime numbers?
(56)

* 13. (Estimate) Claudia's meal cost $7.95. Timo's meal cost $8.95. Estimate the total price for both meals by rounding each amount to the nearest dime before adding.
(61, 62)

* 14. $25 \div 6$
(54)

* 15. $10 \div 9$
(54)

16.
(45)
$$\begin{array}{r} 36.2 \\ 4.7 \\ 15.9 \\ 148.4 \\ 30.5 \\ +\ \ 6.0 \\ \hline \end{array}$$

* 17. 8×503
(59)

* 18. $3w = 36$
(34)

19. $9 \times \$4.63$
(59)

* 20. $8 \times 29¢$
(38)

21.
(28, 52)
$$\begin{array}{r} \$10.00 \\ -\ \$\ 1.73 \\ \hline \end{array}$$

22.
(52)
$$\begin{array}{r} 36{,}428 \\ -\ 27{,}338 \\ \hline \end{array}$$

* 23.
(38)
$$\begin{array}{r} 78 \\ \times\ \ 6 \\ \hline \end{array}$$

* 24. $6 + 5 = 5 + t$
(9)

* 25. $(2 + 3) \times 8 = 5 \times f$
(63)

* 26. $a + 5 = 25 + 5$
(9)

27. $(25 \div 5) + (3 \times 7) - 1$
(63)

* 28. (Explain) Solve the equation below and describe the steps in the order you completed them.
(9, 45)
$$4.7 - (3.6 - 1.7)$$

29. a. Find the perimeter of this rectangle in millimeters.
(20, 42)

    b. Find the area of this rectangle in square centimeters.

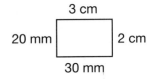

30. Multiple Choice Each angle of this triangle is _____.
(17)
    A acute   **B** right
    C obtuse   **D** straight

\ *California Mathematics Content Standards*

AF 1.0, 1.3 Use parentheses to indicate which operation to perform first when writing expressions containing more than two terms and different operations.

MR 3.0, 3.2 Note the method of deriving the solution and demonstrate a conceptual understanding of the derivation by solving similar problems.

• How Do We Write Expressions?

When we write a math expression, we must write it so that the steps to simplify it are clear. Without clear steps an expression can be simplified incorrectly, which is why we use the order of operations.

Example 1

$$5 \times 4 + 6 \div 2 - 4$$

Unless we know the order of operations, it is not clear which operation should be done first. This expression will be easier to simplify if parentheses are added.

$$(5 \times 4) + (6 \div 2) - 4$$

Now we can see exactly how to simplify this expression.

$$20 + 3 - 4 = 23 - 4 = 19$$

Mathematicians have agreed on an order of operations. Here are some rules arranged in order:

- Do operations in parentheses first.
- Multiply and divide from left to right.
- Add and subtract from left to right.

Evaluate Simplify each expression using the order of operations.

a. $6 + 3 \times (2 - 1) + 9 \div 3$

b. $(7 - 3) + 5 \times 8 - (4 \times 2)$

Example 2

$$(2 \times 3) + ((6 \times 2) \div (3 \times 2)) - 3$$

Sometimes too many parentheses are confusing. Include only those parentheses that are needed or find another way to write the expression.

Using a fraction bar to show the division helps make the order of operations clear.

$$(2 \times 3) + \frac{6 \times 2}{3 \times 2} - 3 = 6 + \frac{12}{6} - 3 = 6 + 2 - 3 = 5$$

Explain Why is it important to be clear when writing math expressions?

Apply Add parentheses two ways to this expression. Then simplify.

$$10 + 6 - (2 \times 4) - 4 \div 2$$

Evaluate Simplify each expression using the order of operations.

c. $(5 \times 1) - ((8 - 3) \times 2 \div (6 + 4)) + 7$

d. $4 \times ((8 + 5) - 6 + (7 \times 0)) - 9$

• Two-Step Equations

California Mathematics Content Standards

AF 1.0, **1.2** Interpret and evaluate mathematical expressions that now use parentheses.

AF 1.0, **1.3** Use parentheses to indicate which operation to perform first when writing expressions containing more than two terms and different operations.

AF 1.0, **1.5** Understand that an equation such as $y = 3x + 5$ is a prescription for determining a second number when a first number is given.

Power Up

facts   Power Up J

count aloud   When we count by fives from 1, we say the numbers 1, 6, 11, 16, and so on. Count by fives from 1 to 51.

mental math   Multiply four numbers in problems **a–c**.

    a. Number Sense: $6 \times 4 \times 10 \times 10$

    b. Number Sense: $3 \times 4 \times 10 \times 10$

    c. Number Sense: $4 \times 5 \times 10 \times 10$

    d. Money: Alex had $10.00. Then he bought a cap for $6.87. How much money does Alex have left?

    e. Time: J'Narra must finish the test by 2:30 p.m. If it is 2:13 p.m., how many minutes does she have left to finish?

    f. Measurement: Five feet is 60 inches. How many inches tall is a person whose height is 5 feet 4 inches?

    g. Estimation: Choose the more reasonable estimate for the width of a computer keyboard: 11 inches or 11 feet.

    h. Calculation: $\sqrt{49} + 6 + 37 + 99$

problem solving   Choose an appropriate problem-solving strategy to solve this problem. Shamel is making lemonade for her lemonade stand. The package of powdered lemonade says that each package makes 1 quart of lemonade. If Shamel wants to make $1\frac{1}{2}$ gallons of lemonade, how many packages of powdered lemonade will she need? Explain how you found your answer.

New Concept

The equation below means "2 times what number equals 7 plus 5?"

$$2n = 7 + 5$$

It takes two steps to solve this equation. The first step is to add 7 and 5 ($7 + 5 = 12$), which gives us this equation:

$$2n = 12$$

The second step is to find n. Since $2 \times 6 = 12$, we know that n is 6.

$$n = \mathbf{6}$$

Verify How can we check the answer?

Example 1

Find m in the following equation: $3m = 4 \cdot 6$

Reading Math

We read this equation as "3 times what number equals 4 times 6?"

A dot is sometimes used between two numbers to indicate multiplication. So $4 \cdot 6$ means "4 times 6." The product of 4 and 6 is 24.

$$3m = 4 \cdot 6$$
$$3m = 24$$

Now we find m. Three times 8 equals 24, so m equals 8.

$$3m = 24$$
$$m = \mathbf{8}$$

Verify How can we check the answer?

Example 2

If $y = 3x + 4$, then what is y when $x = 2$?

Math Symbols

We can show 3 times x in different ways.

$3 \times x$
$3 \cdot x$
$3x$

The equation $y = 3x + 4$ shows us how to find the number that y equals when we know the number x equals.

This equation means, "To find y, multiply x by 3 and then add 4."

In this equation, x is 2, so we multiply 3 times 2 and then add 4.

$y = 3x + 4$	Substitute 2 for x.
$y = (3 \cdot 2) + 4$	Work inside the parentheses first.
$y = 6 + 4$	Add.
$y = 10$	The answer is 10.

When x is 2, $\mathbf{y = 10.}$

Analyze What is y when $x = 3$?

Lesson Practice Find each missing number:

a. $2n = 2 + 8$ **b.** $2 + n = 2 \cdot 8$

c. **Explain** If $y = 2x + 5$, then what is y when $x = 3$? Explain your thinking.

Written Practice Distributed and Integrated

Formulate Write and solve equations for problems **1** and **2**.

1. There were 150 seats in the cafeteria. If 128 seats were filled, how many seats were empty?
(12)

2. Anaya ran 100 meters in 12.14 seconds. Marion ran 100 meters in 11.98 seconds. Marion ran 100 meters how many seconds faster than Anaya?
(16, 45)

3. Forty-two thousand is how much greater than twenty-four thousand?
(27, 52)

4. Keenan bought his lunch Monday through Friday. If each lunch cost $1.25, how much did he spend on lunch for the week?
(39, 59)

5. Find the perimeter and area of this rectangle:
(20, Inv. 3)

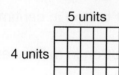

5 units

4 units

6. **Explain** Re'Bekka read 30 pages a day on Monday, Tuesday, and Wednesday. She read 45 pages on Thursday and 26 pages on Friday. How many pages did she read in all? Explain why your answer is reasonable.
(11, 12, 22)

*7. **Evaluate** If $y = 4x + 1$, then what is y when $x = 3$?
(64)

*8. **(Interpret)** This line plot shows the number of times some students
(Inv. 6) ride their bikes in a month. Use this line plot to answer parts **a–c.**

Number of Times Riding a Bike

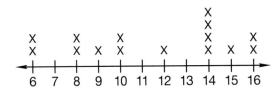

a. How many students were surveyed?

b. What is the mode?

c. Is there an outlier? If yes, what is it?

*9. **(Estimate)** Mr. Anderson bought three packages of chicken. They
(62) weighed 2.36 pounds, 3.75 pounds, and 1.71 pounds. To the nearest
tenth of a pound, how much chicken did Mr. Anderson buy?

10. **(Analyze)** Driving at a highway speed limit of 65 miles per hour, how
(58) far can a truck travel in 3 hours? Make a table to solve this problem.

*11. **(Formulate)** If a biker traveled 24 miles in 4 hours, then the biker
(60) traveled an average of how many miles each hour? Write an equation
to solve this problem.

*12. a. What is the diameter of this shirt button in centimeters?
(18, 42)

b. What is the radius of this shirt button in millimeters?

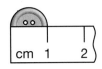

*13. Segment *AB* is 2.7 cm long. Segment *BC* is 4.8 cm long. How long is
(33, 45) segment *AC*? Write a decimal addition equation and find the answer.

14. $8 + $9.48 + 79¢
(28)

15. 5.36 + 2.1 + 0.43
(45)

16. $165.45
(28, 51) + $ 59.47

17.   37,102
(52)   − 18,590

18. $(9 − 5) \times 6 \div 8$
(63)

*19. $1.63 × 4
(59)

*20. 6 × 391
(59)

21. 7 × $2.56
(59)

*22. $3\overline{)19}$
(54)

*23. $9\overline{)40}$
(54)

24. $\dfrac{59}{6}$
(54)

*25. **Represent** Round each number to the given place:
(62)

    a. Round 2.56 to the nearest tenth.

    b. Round $4.35 to the nearest ten cents.

    c. Round $4.35 to the nearest twenty-five cents.

*26. **Analyze** Simplify each expression. Remember to use
(63) the order of operations.

    a. $3 \times 5 + 40 \div 8$ **b.** $64 \div 8 - 6 + 12$

*27. **Formulate** Assume $a = b$. Then use the letters a, b and the number 5
(63) to write an equation that shows "when the same number is added to equal
amounts the sums are equal."

28. **Represent** Write twelve and three tenths as a mixed number and as a
(Inv. 4, 43) decimal number.

*29. **Multiple Choice** Which of these numbers is a factor of both 12
(55) and 20?

    A 3   **B** 4   **C** 5   **D** 6

30. **Represent** Draw a triangle that has one right angle.
(17)

◥ California Mathematics Content Standards

NS 4.0, 4.1 Understand that many whole numbers
break down in different ways (e.g.
$12 = 4 \times 3 = 2 \times 6 = 2 \times 2 \times 3$.)

MR 1.0, 1.2 Determine when and how to break a
problem into simpler parts.

MR 2.0, 2.3 Use a variety of methods, such as
words, numbers, symbols, charts,
graphs, tables, diagrams, and models,
to explain mathematical reasoning.

● Exponents

facts	Power Up J
count aloud	Count by fives from 1 to 51.
mental math	Multiply two numbers ending in zero in problems **a–d.** (Example: 30×40 equals 3×10 times 4×10. We rearrange the factors to get $3 \times 4 \times 10 \times 10$, which is 1200.)

    a. Number Sense: 40×40

    b. Number Sense: 30×50

    c. Number Sense: 60×70

    d. Number Sense: 40×50

    e. Powers/Roots: $2^2 + 2$

    f. Money: $\$6.48 + \$2.39$

    g. Estimation: Each bottled water costs 99¢. If Ms. Hathcoat buys 1 bottle for each of her 24 students, about how much money will she spend?

    h. Calculation: $\sqrt{64} - 6 + 37 + 61$

problem solving

Choose an appropriate problem-solving strategy to solve this problem. Jamisha paid a dollar for an item that cost 44¢. If she got back four coins in change, what should the four coins have been?

New Concept

To find the product of three numbers, we first multiply two of the numbers. Then we multiply the answer we get by the third number. To multiply four numbers, we must multiply once more. In any multiplication we continue the process until no factors remain.

Example 1

Multiply: $3 \times 4 \times 5$

First we multiply two of the numbers to get a product. Then we multiply that product by the third number. If we multiply 3 by 4 first, we get 12. Then we multiply 12 by 5 and get 60.

STEP 1	STEP 2
3	12
$\times$ 4	$\times$ 5
12	**60**

It does not matter which two numbers we multiply first. If we multiply 5 by 4 first, we get 20. Then we multiply 20 by 3 and again get 60.

STEP 1	STEP 2	
5	20	
$\times$ 4	$\times$ 3	
20	**60**	← same answer

The order of the multiplications does not matter because of the Commutative Property of Multiplication, which we studied in Lesson 23.

Example 2

Multiply: $4 \times 5 \times 10 \times 10$

We may perform this multiplication mentally. If we first multiply 4 by 5, we get 20. Then we multiply 20 by 10 to get 200. Finally we multiply 200 by 10 and find that the product is **2000.**

(**Formulate**) What is another combination of factors that has a product of 2000?

Sometimes when we simplify an expression that includes exponents, we will multiply more than two factors.

An **exponent** is a number that shows how many times another number (the **base**) is to be used as a factor. An exponent is written above and to the right of the base.

$$\text{base} \longrightarrow 5^2 \longleftarrow \text{exponent}$$

$$5^2 \text{ means } 5 \times 5.$$
$$5^2 \text{ equals } 25.$$

If the exponent is 2, we say "squared" for the exponent. So 5^2 is read as "five squared." If the exponent is 3, we say "cubed" for the exponent. So the **exponential expression** 2^3 is read as "two cubed."

Example 3

Simplify: $5^2 + 2^3$

We will add five squared and two cubed. We find the values of 5^2 and 2^3 before adding.

5^2 means 5×5, which is 25.

2^3 means $2 \times 2 \times 2$, which is 8.

Now we add 25 and 8.

$$25 + 8 = \textbf{33}$$

Example 4

Rewrite this expression using exponents:

$$\textbf{5} \times \textbf{5} \times \textbf{5}$$

Five is used as a factor three times, so the exponent is 3.

$$\textbf{5}^3$$

(**Evaluate**) Why do we write $5 \times 5 \times 5$ as 5^3 instead of 5×3?

(**Lesson Practice**) Simplify:

a. $2 \times 3 \times 4$ **b.** $3 \times 4 \times 10$

c. 8^2 **d.** 3^3

e. $10^2 - 6^2$ **f.** $3^2 - 2^3$

g. Rewrite this expression using exponents:

$$4 \times 4 \times 4$$

(**Written Practice**) *Distributed and Integrated*

---

*1. A rectangular wall is covered with square tiles. The wall is 4 tiles long and
(Inv. 3) 3 tiles wide. In all, how many tiles are on the wall?

2. There were two hundred sixty seats in the movie theater. All but forty-three
(12) seats were occupied. How many seats were occupied?

3. At the grand opening of a specialty food store, five coupons were
(39, 59) given to each customer. One hundred fifteen customers attended the
grand opening. How many coupons were given to those customers
altogether?

4. (**Analyze**) What is the value of 5 pennies, 3 dimes, 2 quarters, and
(28) 3 nickels?

*5. **(Evaluate)** If $y = 2x + 3$, then what is y when $x = 4$?
(64)

*6. **(Interpret)** Use this data to answer parts **a** and **b.**
(Inv. 6)
$$20, 30, 50, 80, 40, 10, 90$$
  a. What is the mode?

  b. What is the median?

*7. **(Represent)** Round each number to the given place.
(62)
  a. Round 3.12 to the nearest tenth.

  b. Round $7.55 to the nearest ten cents.

*8. a. The line segment shown below is how many centimeters long?
(42)
  b. The segment is how many millimeters long?

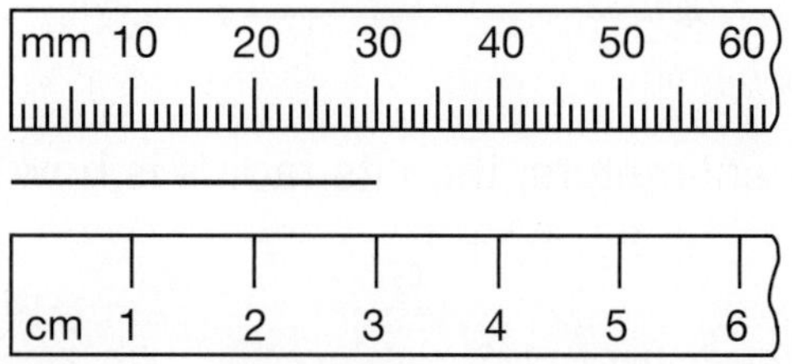

9. The first four multiples of 9 are 9, 18, 27, and 36. What are the first four
(55) multiples of 90?

10. **(Represent)** Compare: $\frac{2}{3} \bigcirc \frac{2}{5}$. Draw and shade two equal rectangles
(57) to show the comparison.

11. Badu can ride her bike an average of 12 miles per hour. At that
(58) rate, how many miles could she ride in 4 hours? Make a table to
solve this problem.

12. $375.48
(28, 51) + $536.70

13.   367,419
(51) + 90,852

14.   42.3
(45)   57.1
  28.9
  96.4
 + 38.0

15. $20.00
(28, 52) − $19.39

16.   310,419
(52) − 250,527

17. $\$6.08$
(59)
$\times\ \ \ \ 7$

18. 86
(38)
$\times\ \ 4$

19. $59¢$
(38)
$\times\ \ 8$

* **20.** $3\overline{)23}$
(54)

* **21.** $8\overline{)30}$
(54)

* **22.** $5\overline{)33}$
(54)

* **23.** $(8 \times 2) \div (2 \cdot 2)$
(63)

* **24.** $\sqrt{36} + 4^2 + 10^2$
(Inv. 3, 65)

25. $9 + m = 27 + 72$
(64)

26. $6n = 4 \cdot 12$
(64)

* **27.** (**Explain**) If $p = r$ is the equation below true? How do you know?
(63)
$$p \cdot 100 = r \cdot 100$$

28. (**Model**) Use an inch ruler to find the lengths of segments *AB*, *BC*,
(33, 42) and *AC*.

A B C

* **29.** If the diameter of a coin is 2 centimeters, then its radius is how many
(18, 42) millimeters?

* **30.** (**Estimate**) From 7 a.m. until noon, the employees in a customer
(61) service department received 47 phone calls. What is a reasonable estimate of the number of calls that were received each hour? Explain how you found your answer.

California Mathematics Content Standards

NS 2.0, 2.2 Round two-place decimals to one decimal or the nearest whole number and judge the reasonableness of the rounded answer.

MG 1.0, 1.1 Measure the area of rectangular shapes by using appropriate units, such as square centimeter (cm²), square meter (m²), square kilometer (km²), square inch (in²), square yard (yd²), or square mile (mi²).

MG 1.0, 1.4 Understand and use formulas to solve problems involving perimeters and areas of rectangles and squares. Use those formulas to find the areas of more complex figures by dividing the figures into basic shapes.

• Area of a Rectangle

facts	Power Up J
count aloud	Count down by fives from 51 to 1.
mental math	Multiply three numbers, including numbers ending in zero, in problems **a–c**.

    a. Number Sense: $3 \times 10 \times 20$

    b. Number Sense: $4 \times 20 \times 30$

    c. Number Sense: $3 \times 40 \times 10$

    d. Powers/Roots: $2^2 + 5^2$

    e. Geometry: Altogether, how many sides do 3 hexagons have?

    f. Money: Logan owes $10.00 for his club dues. He has $9.24. How much more money does Logan need?

    g. Estimation: Liev wants to buy 6 stickers that each cost 21¢. Liev has $1.15. Does he have enough money to buy 6 stickers?

    h. Calculation[1]**:** $\sqrt{16}$, × 2, × 2, + 4, × 2

problem solving	Choose an appropriate problem-solving strategy to solve this problem. Dasha plans to use only four different colored pencils to color the states on a United States map. She has five different colored pencils from which to choose—red, orange, yellow, green, and blue. What are the combinations of four colors Dasha can choose? (There are five combinations.)

---

[1] As a shorthand, we will use commas to separate operations to be performed sequentially from left to right. In this case, $\sqrt{16} = 4$, then $4 \times 2 = 8$, then $8 \times 2 = 16$, then $16 + 4 = 20$, then $20 \times 2 = 40$. The answer is 40.

We have learned to find the perimeter and area of rectangles. We know that the distance around a figure is its perimeter and the number of square units that cover a figure is its area.

Today we are going to learn a formula to find the area of a rectangle. We know that we can multiply 3 by 6 to find the total number of square centimeters that cover this figure.

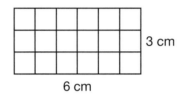

6 × 3 = 18 square cm

We multiplied the length times the width to find the area. So, the formula for finding the area is Area = length × width or $A = l \times w$.

$$A = l \times w$$
$$= 6 \text{ cm} \times 3 \text{ cm}$$
$$= (6 \times 3) \times (\text{cm} \times \text{cm})$$
$$= 18 \text{ sq. cm or } 18 \text{ cm}^2$$

We read 18 cm² as 18 square centimeters.

Example 1

Robin has a box that is 8 in. wide and 10 in. long.

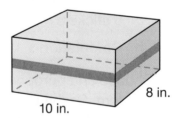

a. She wants to paint a stripe all around the box. How long will the stripe be?

b. She wants to use fabric inside the box to cover bottom. How much fabric does she need?

a. The stripe will be painted around the box. We can use the perimeter formula to find the distance around the box.

$$P = 2l + 2w$$
$$= (2 \times 10) + (2 \times 8)$$
$$= 20 + 16$$
$$= 36 \text{ in.}$$

The stripe will be **36 in.** long.

b. The fabric liner will cover the bottom of the box. We can use the area formula to find the area of the bottom of the box.

$$A = l \times w$$
$$= 10 \times 8$$
$$= 80$$
$$= 80 \text{ in}^2$$

Robin needs **80 in²** of fabric.

Discuss Why did we write the amount of fabric as 80 in² and not 80 in.?

Example 2

Use the formula for the area of a square to find the area of this square.

The formula for the area of a square is $A = s^2$. The length of each side is 5 in. Replace the "s" in the formula with 5 in.

$$A = (5 \text{ in.})^2$$

Multiplying 5 in. $\times$ 5 in., we find the area of the square is 25 sq. in. We can write the answer **25 in²** or **25 sq. in.**

5 in.

Example 3

Michelle wants to have her driveway repaved. The driveway is 8.25 ft by 42.75 ft. Estimate the area of the driveway.

To estimate, we will first we round each measurement to the nearest whole number.

8.25 ft rounds to 8 ft

42.75 ft rounds to 43 ft

Now we can use the area formula.

$$A = l \times w$$
$$= 43 \text{ ft} \times 8 \text{ ft}$$
$$= 344 \text{ ft}^2$$

The area of the driveway is **344 ft²**.

Lesson Practice

a. Matthew's kitchen is 9 ft wide and 15 ft long. What is the perimeter and area of the room?

b. What is the area of this rectangle rounded to the nearest whole number?

7.8 m

29.5 m

Written Practice *Distributed and Integrated*

1. Christie's car travels 18 miles on each gallon of gas. How many miles can it travel on 10 gallons of gas?
(58)

*2. **Analyze** Alejandro mowed a yard that was 50 feet wide. Each time he pushed the mower along the length of the yard, he mowed a path 24 inches wide. To mow the entire yard, how many times did Alejandro need to push the mower along the length of the yard?
(20, 42, 53)

*3. A gift of $60 is to be divided equally among 6 children. What amount of money will each child receive?
(53)

4. Soccer practice lasts for an hour and a half. If practice starts at 3:15 p.m., at what time does it end?
(13)

*5. a. Round 4.37 to the nearest tenth.
(46, 62)

b. Round 4.37 to the nearest whole number.

c. Round $8.34 to the nearest ten cents.

d. Round $8.34 to the nearest dollar.

6. Find the perimeter and area of the rectangle at right.
(66)

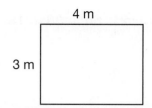

4 m

3 m

*7. (Estimate) This key is 60 mm long. The key is how many centimeters long?
(42)

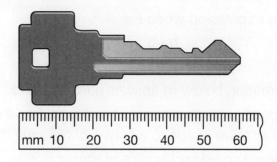

8. According to this calendar, the year 1902 began on what day of the week?
(RF12)

DECEMBER 1901						
S	M	T	W	T	F	S
1	2	3	4	5	6	7
8	9	10	11	12	13	14
15	16	17	18	19	20	21
22	23	24	25	26	27	28
29	30	31				

*9. (Estimate) A sheet of paper is 8.5 in. wide and 11 in. long. Estimate the area of the paper. Explain your thinking.
(Inv. 3, 61)

10. A meter equals 100 centimeters. If each side of a square is 1 meter long, then what is the perimeter of the square in centimeters?
(20, 42)

11. List the first four multiples of 70.
(55)

12. $1.68 + 32¢ + $6.37 + $5
(28)

13. 4.3 + 2.4 + 0.8 + 6.7
(45)

14. (Explain) Find $10 − ($6.46 + $2.17). Describe the steps you used.
(9, 28, 52)

15. 5 × 4 × 5   **16.** 359 × 7   **17.** 5 × 74
(65)   (59)   (38)

18. 4)30   **19.** 5)43   **20.** 8)76
(54)   (54)   (54)

*21. $6n = 30 + 18$
(64)

*22. $3^3 + 2^3$
(65)

*23. If $y = 2x + 8$, then what is y when $x = 6$?
(64)

Connect Simplify each expression. Remember to use the order of operations.

*24. $8 \times 5 + 45 \div 9$
(63)

*25. $49 \div 7 - 5 + 9$
(63)

*26. **Evaluate** Simplify each expression when $v = 4$.
(63)
  a. $36 - 12 \div v$ **b.** $6 + (8 - v) \times 10$

*27. **Analyze** Use the information below to answer parts **a** and **b.**
(12)

  Kamili scored two goals when her soccer team won 5 to 4 on November 3. To make the playoffs, her team needs to win two of the next three games.

  a. How many goals were scored by Kamili's teammates?

  b. Kamili's team has won four games and lost three games. Altogether, how many games does Kamili's team need to win to make the playoffs?

28. a. **Classify** Angles C and D of this polygon are right angles. Which angle appears to be an obtuse angle?
(17)

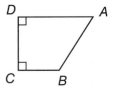

  b. **Classify** Which segments are perpendicular?

  c. **Classify** Which segments are parallel?

*29. **Multiple Choice** Which of the following is *not* a hexagon?
(50)

*30. **Represent** The average weights of some animals are shown in the table. Make a bar graph to display the data.
(Inv. 5)

Average Weights of Animals

Animal	Weight (in pounds)
Domestic Rabbit	8
Otter	13
Ringtail Monkey	6
Chicken	7

✎ *California Mathematics Content Standards*

NS 1.0, 1.5 Explain different interpretations of fractions, for example, parts of a whole, parts of a set, and division of whole numbers by whole numbers; explain equivalents of fractions (see Standard 4.0).

NS 1.0, 1.7 Write the fraction represented by a drawing of parts of a figure; represent a given fraction by using drawings; and relate a fraction to a simple decimal on a number line.

• Remaining Fractions

facts	Power Up I
count aloud	When we count by fives from 2, we say the numbers 2, 7, 12, 17, and so on. Count by fives from 2 to 52.
mental math	Multiply numbers ending in two zeros by numbers ending in one zero in problems **a–c**.

    a. Number Sense: 200×10

    b. Number Sense: 300×20

    c. Number Sense: 400×50

    d. Fractional Part: $\frac{1}{2}$ of \$10

    e. Fractional Part: $\frac{1}{4}$ of \$10

    f. Fractional Part: $\frac{1}{10}$ of \$10

    g. Estimation: Estimate the total cost of two items priced at \$3.88 each and one item priced at \$5.98.

    h. Calculation: 4^2, $+ 34$, $+ 72$, $- 24$

problem solving

Choose an appropriate problem-solving strategy to solve this problem. Mathea exercised for half of an hour. For half of her exercise time, she was running. For how many minutes was Mathea exercising? For how many minutes was she running?

New Concept

The whole circle in Example 1 on the following page has a shaded portion and an unshaded portion. If we know the size of one portion of a whole, then we can figure out the size of the other portion.

Example 1

a. **What fraction of the circle is shaded?**

b. **What fraction of the circle is not shaded?**

We see that the whole circle has been divided into eight equal parts. Three of the parts are shaded, so five of the parts are not shaded.

a. The fraction that is shaded is $\frac{3}{8}$.

b. The fraction that is not shaded is $\frac{5}{8}$.

Represent Compare the shaded part to the part not shaded using >, <, or =.

Example 2

The quesadilla was cut into eight equal slices. After Willis, Hunter, and Svelita each took a slice, what fraction was left?

The whole quesadilla was cut into eight equal parts. Since three of the eight parts were taken, five of the eight parts remained. The fraction that was left was $\frac{5}{8}$.

Example 3

Two fifths of the crowd cheered. What fraction of the crowd did not cheer?

We think of the crowd as though it were divided into five equal parts. We are told that two of the five parts cheered. So there were three parts that did not cheer. The fraction of the crowd that did not cheer was $\frac{3}{5}$.

Lesson Practice

a. **What fraction of this rectangle is not shaded?**

b. **Three fifths of the race was over. What fraction of the race was left?**

Written Practice   *Distributed and Integrated*

* **1.** Two thirds of the pencils are sharpened. What fraction of the pencils
(67)   are *not* sharpened? Draw a picture to solve the problem.

*2. If $y = 5x + 10$, then what is y when $x = 5$?
(64)

*3. Use this information to answer parts **a–c:**
(28, 39, 53)

> *Thirty students are going on a field trip. Each car can hold five students. The field trip will cost each student $5.*

a. How many cars are needed for the field trip?

b. Altogether, how much money will be needed?

c. Don has saved $3.25. How much more does he need to go on the field trip?

4. **(Analyze)** During the summer the swim team practiced $3\frac{1}{2}$ hours a day.
(13) If practice started at 6:30 a.m., at what time did it end if there were no breaks?

*5. a. Round 5.26 to the nearest tenth.
(46, 62)

b. Round 5.26 to the nearest whole number.

c. Round $10.65 to the nearest ten cents.

d. Round $10.65 to the nearest dollar.

6. A mile is five thousand, two hundred eighty feet. The Golden Gate
(16, 42, 52) Bridge is four thousand, two hundred feet long. The Golden Gate Bridge is how many feet less than 1 mile long?

7. **Multiple Choice** Which of these numbers is *not* a multiple of 90?
(55)
A 45 **B** 180 **C** 270 **D** 360

8. What number is halfway between 300 and 400?
(Inv. 2)

9. 37.56 − 4.2 **10.** 4.2 + 3.5 + 0.25 + 4.0
(45) (45)

11. Each side of a regular polygon has the same length. A
(20, 42) regular hexagon is shown to the right. How many millimeters is the perimeter of this hexagon?

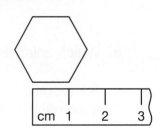

12. $\sqrt{25} \times m = 45$ **13.** $z − 476 = 325$ **14.** $6a = 12 + 6$
(Inv. 3, 64) (6, 8) (64)

15. $100.00
(28, 52)   − $ 31.53

16. 251,546
(52)   − 37,156

17. n
(8, 10)   + 423
          ‾‾‾‾‾
           618

18. $3.46
(59) ×   7

19.   96
(38)   ×   3

20. $0.59
(59) ×   8

* 21. 7)‾65
(54)

* 22. 5)‾38
(54)

* 23. 3)‾17
(54)

(**Analyze**) Simplify. Use the order of operations.

* 24. $30 + 10 \times 5$
(63)

* 25. $64 - 8 \div 4$
(63)

* 26. $56 \div 7 \times 5$
(65)

* 27. (**Connect**) Segment *AB* is 2.3 cm long. Segment *BC* is 3.5 cm long.
(33, 42, 45) How long is segment *AC*? Write a decimal addition problem and find
the answer.

A   B   C

* 28. Jennie wants to tile the top of a box that is 6 in. wide and 8 in. long.
(66)   How many one-inch square tiles does she need?

29. (**Estimate**) Using rounding, which numbers would you choose to
(61)   estimate the product of 2 × 65? Explain your reasoning.

* 30. (**Interpret**) This pictograph shows the maximum speeds that
(Inv. 5, 61) animals can run for a short distance. Use the pictograph to answer the
questions that follow.

Animal	Maximum Speed (in miles per hour)
Wart hog	🐎 🐎 🐎
Wild turkey	🐎 🐎
Lion	🐎 🐎 🐎 🐎 🐎
Elephant	🐎 🐎 🐎
Zebra	🐎 🐎 🐎 🐎

Key: 🐎 = 10 miles per hour

a. Which animals can run at a speed of at least 30 miles per hour?

b. A squirrel can run at a maximum speed of 12 miles per hour. About
how many times greater is the maximum speed of a lion? Explain.

LESSON

68

California Mathematics Content Standards

NS 1.0, **1.3** Round whole numbers through the millions to the nearest ten, hundred, thousand, ten thousand, or hundred thousand.

AF 1.0, 1.1 Use letters, boxes, or other symbols to stand for any number in simple expressions or equations (e.g., demonstrate an understanding and the use of the concept of a variable).

MR 1.0, 1.1 Analyze problems by identifying relationships, distinguishing relevant from irrelevant information, sequencing and prioritizing information, and observing patterns.

• Division with Two-Digit Answers, Part 1

Power Up

facts	Power Up I
count aloud	Count down by fives from 52 to 2.
mental math	**a. Number Sense:** $10 \times 20 \times 30$
	b. Number Sense: 250×10
	c. Money: Shatavia had $5.00. Then she spent $3.79. How much did she have left?
	d. Money: Tan bought a scorebook for $6.48 and a whistle for $2.84. How much did he spend?
	e. Geometry: What is the perimeter of a square with 9-inch sides? Express your answer in feet.
	f. Time: How many years is 1 century plus 4 decades?
	g. Estimation: Estimate 193×5 by rounding 193 to the nearest hundred and then multiplying.
	h. Calculation: $18 \div 9$, $\times 6$, $\times 6$

problem solving

Choose an appropriate problem-solving strategy to solve this problem. Stephanie solved an addition problem and then erased some of the digits from the problem. She gave it to Ian as a problem-solving exercise. Copy Stephanie's problem on your paper, and find the missing digits for Ian.

$$
\begin{array}{r}
7\_6 \\
+ \ \_4\_ \\
\hline
\_45
\end{array}
$$

New Concept

In this lesson we will learn a pencil-and-paper method for dividing a two-digit number by a one-digit number. We will demonstrate the method as we solve the problem on the next page.

The seventy-eight fifth-graders at Washington School will be divided equally among three classrooms. How many students will be in each room?

There are three numbers in this "equal groups" problem: the total number of students, the number of classrooms, and the number of students in each classroom.

Formula:

Number **of** groups × Number **in each** group = Total

Problem:

3 classrooms × n students in each classroom = 78 students

> **Reading Math**
>
> We can write the related equation $78 \div 3 = n$ to represent this problem.

To find the number of students in each classroom, we divide 78 by 3.

$$3\overline{)78}$$

> **Thinking Skills**
>
> **Discuss**
>
> Why do we write the first digit of the quotient in the tens place?

For the first step we ignore the 8 and divide 7 tens by 3. We write "2" above the 7. Then we multiply 2 by 3 and write "6" below the 7 tens. Then we subtract and write "1."

$$\begin{array}{r} 2 \\ 3\overline{)78} \\ \underline{6} \\ 1 \end{array}$$

Next we "bring down" the 8, as shown here. Together, the 1 ten and 8 form 18 ones.

$$\begin{array}{r} 2 \\ 3\overline{)78} \\ \underline{6}\downarrow \\ 18 \end{array}$$

> **Thinking Skills**
>
> **Verify**
>
> Why do we write the second digit of the quotient in the ones place?

Now we divide 18 by 3 and get 6. We write the 6 above the 8 in 78. Then we multiply 6 by 3 and write "18" below the 18.

$$\begin{array}{r} 26 \\ 3\overline{)78} \\ \underline{6} \\ 18 \\ \underline{18} \\ 0 \end{array}$$

We subtract and find that the remainder is zero. This means that if the students are divided equally among the classrooms, there will be 26 students in each classroom.

$$78 \div 3 = 26$$

Since multiplication and division are inverse operations, we may arrange these three numbers to form a related multiplication equation.

$$3 \times 26 = 78$$

We can multiply 26 by 3 to check our work.

$$\begin{array}{r} \overset{1}{26} \\ \times \ \ 3 \\ \hline 78 \end{array} \quad \text{check}$$

Example 1

An 87-acre field is divided into 3 equal parts. A different crop will be planted in each part. How many acres is one part of the field?

For the first step we ignore the 7. We divide 8 tens by 3, multiply, and then subtract. Next we bring down the 7 to form 27 ones. Now we divide 27 by 3, multiply, and subtract again.

$$\begin{array}{r} 29 \\ 3\overline{)87} \\ 6\downarrow \\ \hline 27 \\ 27 \\ \hline 0 \end{array}$$

The remainder is zero, so we see that one part of the field is **29 acres.**

Now we multiply 29 by 3 to check our work. If the product is 87, we can be confident that our division was correct.

$$\begin{array}{r} \overset{2}{29} \\ \times \ \ 3 \\ \hline 87 \end{array} \quad \text{check}$$

Notice that there is no remainder when 87 is divided by 3. That is because 87 is a multiple of 3. We cannot identify the multiples of 3 by looking at the last digit, because the multiples of 3 can end with any digit. However, adding the digits of a number can tell us whether a number is a multiple of 3. If the sum is a multiple of 3, then so is the number. For example, adding the digits in 87 gives us 15 (8 + 7 = 15). Since 15 is a multiple of 3, we know that 87 is a multiple of 3.

Example 2

Four students can sit in each row of seats in a school bus. Thirty-eight students are getting on the bus. If each student sits in the first available seat, what is a reasonable estimate of the number of rows of seats that will be filled?

We are asked for a reasonable estimate, so we don't need to find an exact answer. We can round 38 to 40 and divide by 4. We find that a reasonable estimate of the number of rows that will be filled is **10.**

Example 3

Which of these numbers can be divided by 3 with no remainder?

A 56   **B** 64   **C** 45   **D** 73

We add the digits of each number:

A $5 + 6 = 11$ **B** $6 + 4 = 10$ **C** $4 + 5 = 9$ **D** $7 + 3 = 10$

Of the numbers 11, 10, and 9, only 9 is a multiple of 3. So the only choice that can be divided by 3 with no remainder is **45.**

Lesson Practice   Divide:

a. $3\overline{)51}$ **b.** $4\overline{)52}$ **c.** $5\overline{)75}$

d. $3\overline{)72}$ **e.** $4\overline{)96}$ **f.** $2\overline{)74}$

g. **Connect** Find the missing factor in this equation: $3n = 45$

h. **Multiple Choice** Which of these numbers can be divided by 3 with no remainder? How do you know?

A 75   **B** 76

C 77   **D** 79

i. Each row of desks in a classroom can seat six students. Twenty-nine students are entering the classroom. If each student sits in the first available seat, what is a reasonable estimate of the number of rows of seats that will be filled? Explain your answer.

Written Practice   *Distributed and Integrated*

1. Michael volunteered for sixty-two hours last semester. Milagro
(11, 12)   volunteered for seven hours. Mitsu and Michelle each volunteered for twelve hours. Altogether, how many hours did they volunteer?

*2. The Matterhorn is fourteen thousand, six hundred ninety-one feet high.
(16, 52) Mont Blanc is fifteen thousand, seven hundred seventy-one feet high.
How much taller is Mont Blanc than the Matterhorn?

3. There are 25 squares on a bingo card. How many squares are on
(38, 39) 4 bingo cards?

*4. **Analyze** Ninety-six books were placed on 4 shelves so that the same
(53, 68) number of books were on each shelf. How many books were on each
shelf?

5. How many years is ten centuries? (*Hint:* A century is 100 years)
(RF12, 37)

6. **Estimate** A package of Jose's favorite trading cards costs $1.75.
(61) What is a reasonable estimate of the number of packages Jose could
purchase with $10.00? Explain your answer.

*7. Two fifths of the bottle is empty. What fraction of the bottle is
(67) *not* empty? Draw a picture to solve the problem.

*8. If $y = 3x + 8$, then what is y when $x = 1$?
(64)

*9. **Connect** Simplify each expression. Use the order of operations.
(63)
    a. $32 + 32 \div 4 \times 7$ **b.** $27 - 12 \times 2 + 7$

    c. How could we place parentheses in the expressions for **a** and **b** so it clear how
to perform the order of operations?

10. a. What is the perimeter of the rectangle shown at right?
(20, Inv. 3)

    b. How many 1-inch squares would be needed to cover
this rectangle?

6 in.

3 in.

11. **Predict** How many millimeters are equal to 10 centimeters? Use the
(25, 42) table to decide.

Millimeters	10	20	30	40	50
Centimeters	1	2	3	4	5

*12. **Analyze** Mrs. Noh has an herb garden that is 4 ft wide and 9 ft long.
(31, 66)
    a. What is the area of her garden?

b. If she doubled the area of her garden, what could the new dimensions be?

13. $6.15 − ($0.57 + $1.20)
(28)

14. 43,160 − 8459
(52)

15. 8 × 8 × 8
(65)

16. $3.54 × 6
(59)

17. 8 × 57
(38)

18. 704 × 9
(59)

19. $9\overline{)87}$
(54)

20. $7\overline{)32}$
(54)

21. $5\overline{)48}$
(54)

*22. 96 ÷ 3
(68)

*23. $\dfrac{85}{5}$
(68)

24. 96 ÷ 8
(68)

*25. $\sqrt{36} + n = 6^2$
(Inv. 3, 65)

26. 462 − y = 205
(8, 10)

27. 50 = 5r
(34)

28. **Conclude** Find the next number in this counting sequence:
(37, 55)

..., 60, 120, 180, _____, ...

*29. **Explain** Sierra's arm is 20 inches long. If Sierra swings her arm in a
(18) circle, what will be the diameter of the circle? Explain your answer.

30. **Multiple Choice** Which of these numbers is a prime number?
(56)
    A 1   **B** 2   **C** 4   **D** 9

• Division with Two-Digit Answers, Part 2

✎ *California Mathematics Content Standards*

NS 3.0 3.2 Demonstrate an understanding of, and the ability to use, standard algorithms for multiplying a multidigit number by a two-digit number and for dividing a multidigit number by a one-digit number; use relationships between them to simplify computations and to check results.

NS 3.0, 3.4 Solve problems involving division of multidigit numbers by one-digit numbers.

AF 1.0, 1.1 Use letters, boxes, or other symbols to stand for any number in simple expressions or equations (e.g., demonstrate an understanding and the use of the concept of a variable).

facts	Power Up I
count aloud	Count down by threes from 60 to 3.
mental math	**a. Number Sense:** $12 \times 2 \times 10$
	b. Number Sense: $20 \times 20 \times 20$
	c. Number Sense: $56 + 9 + 120$
	d. Fractional Parts: What is $\frac{1}{2}$ of 60?
	e. Measurement: Six feet is 72 inches. How many inches tall is a person whose height is 5 feet 11 inches?
	f. Measurement: The airplane is 5500 feet above the ground. Is that height greater than or less than 1 mile?
	g. Estimation: Xavier can read about 30 pages in one hour. If Xavier must read 58 pages, about how long will it take him? (Round your answer to the nearest hour.)
	h. Calculation: $6^2, - 18, \div 9, \times 50$
problem solving	Choose an appropriate problem-solving strategy to solve this problem. The parking lot charged $1.50 for the first hour and 75¢ for each additional hour. Harold parked the car in the lot from 11:00 a.m. to 3 p.m. How much money did he have to pay? Explain how you found your answer.

New Concept

We solve the following problem by dividing:

On a three day bike trip Hans rode 234 kilometers. Hans rode an average of how many kilometers each day?

We find the answer by dividing 234 by 3.

$$3\overline{)234}$$

To perform the division, we begin by dividing $3\overline{)23}$. We write "7 tens" above the 3 tens of 23. Then we multiply and subtract.

$$\begin{array}{r} 7 \\ 3\overline{)234} \\ \underline{21} \\ 2 \end{array}$$

Next we bring down the 4.

$$\begin{array}{r} 7 \\ 3\overline{)234} \\ \underline{21}\downarrow \\ 24 \end{array}$$

Now we divide 24 by 3. We write "8" above the 4 ones. Then we multiply and finish by subtracting.

$$\begin{array}{r} 78 \\ 3\overline{)234} \\ \underline{21} \\ 24 \\ \underline{24} \\ 0 \end{array}$$

We find that Hans rode an average of 78 kilometers each day.

We can check our work by multiplying the quotient, 78, by the divisor, 3. If the product is 234, then our division answer is correct.

$$\begin{array}{r} 78 \\ \times\ 3 \\ \hline 234 \end{array} \quad \text{check}$$

Thinking Skills

(Discuss)

Why do we write the first digit of the quotient in the tens place?

Thinking Skills

(Verify)

Why do we write the second digit of the quotient in the ones place?

Example 1

On a 9-day bike trip through the Rocky Mountains, Vera and her companions rode 468 miles. They rode an average of how many miles per day?

Vera and her companions probably rode different distances each day. By dividing 468 miles by 9, we find how far they traveled if they rode the same distance each day. This is called the *average distance*. We begin by finding $9\overline{)46}$. We write "5" above the 6 in 46. Then we multiply and subtract.

$$\begin{array}{r} 5 \\ 9\overline{)468} \\ \underline{45} \\ 1 \end{array}$$

Next we bring down the 8. Now we divide 18 by 9.

$$
\begin{array}{r}
52 \\
9\overline{)468} \\
45\downarrow \\
\overline{18} \\
18 \\
\overline{0}
\end{array}
$$

We find that they rode an average of **52 miles** per day.

We check the division by multiplying 52 by 9, and we look for 468 as the answer.

$$
\begin{array}{r}
1 \\
52 \\
\times\ 9 \\
\overline{468} \quad \text{check}
\end{array}
$$

(**Connect**) Why can we use multiplication to check a division problem?

Notice in Example 2 that there is no remainder when 468 is divided by 9. That is because 468 is a multiple of 9. Just as we identified multiples of 3 by adding the digits of a number, we can identify multiples of 9 by adding the digits of a number. For the number 468, we have

$$4 + 6 + 8 = 18$$

The sum 18 is a multiple of 9, so 468 is a multiple of 9.

Example 2

Which of these numbers is a multiple of 9?

A 123　　　　**B** 234　　　　**C** 345　　　　**D** 456

We add the digits of each number:

A $1 + 2 + 3 = 6$　　　　**B** $2 + 3 + 4 = 9$

C $3 + 4 + 5 = 12$　　　　**D** $4 + 5 + 6 = 15$

The sums 6, 9, and 12 are all multiples of 3, but only 9 is a multiple of 9. Therefore, only **234** is a multiple of 9 and can be divided by 9 without a remainder.

Lesson Practice　In the division fact $32 \div 8 = 4$,

　　　a. what number is the divisor?

　　　b. what number is the dividend?

　　　c. what number is the quotient?

Divide:

d. $3\overline{)144}$ **e.** $4\overline{)144}$ **f.** $6\overline{)144}$

g. $225 \div 5$ **h.** $455 \div 7$ **i.** $200 \div 8$

j. Multiple Choice Which of these numbers can be divided by 9 without a remainder? How do you know?

    A 288   **B** 377   **C** 466   **D** 555

k. Find the missing factor in this equation:

$$5m = 125$$

Written Practice
Distributed and Integrated

1. Pears cost 59¢ per pound. How much would 4 pounds of pears cost?
(39)

2. Find the perimeter and area of this rectangle:
(66)

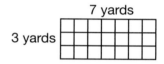

*3. **Connect** There were three hundred sixty books on the floor. Frankie
(69) put half of the books on a table.

    a. How many books did Frankie put on the table?

    b. How many books were still on the floor?

*4. If $y = 2x + 7$, then what is y when $x = 9$?
(64)

*5. **Represent** To what decimal number is the arrow pointing? What
(43) mixed number is this?

6. Estimate Two hundred seventy-two students attend one elementary
(61) school in a city. Three hundred nineteen students attend another elementary school. Estimate the total number of students attending those schools by rounding the number of students attending each school to the nearest hundred before adding.

*7. Five sixths of the rolls have been sold. What fraction of the rolls have
_(67)_ *not* been sold? Draw a picture to solve the problem.

*8. (**Connect**) Simplify each expression. Use for the order of operations.
_(63)_
a. 25 − (18 ÷ 3) + 7   **b.** 30 + (40 ÷ 2) − 5

*9. (**Evaluate**) (44 ÷ 2) + (2 × 0) − *t* when *t* = 10.
_(63)_

*10. (**Interpret**) Use the survey information in the bar graph below to
(Inv. 5,
Inv. 6) answer parts **a–c.**

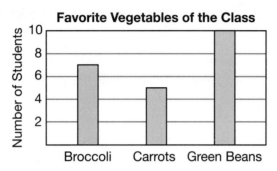

a. How many students were surveyed?

b. Carrots are the favorite vegetable of how many students?

c. Is it correct to say that most of the students chose green beans
as their favorite vegetable? Why or why not?

11. (**Represent**) The 8 a.m. temperature was −5 degrees Fahrenheit.
_(21)_ By 3 p.m., the temperature had increased 10 degrees. What was the
3 p.m. temperature?

*12. (**Represent**) Round each number to the given place:
_(62)_
a. Round 10.37 to the nearest tenth.

b. Round $25.25 to the nearest ten cents.

13. $86.47
_(51)_ + $47.98

14. 36.7
_(45)_ − 18.5

15. 2358
_(51)_ 4715
 317
 2103
 + 62

*16. 3)93
_(68)_

*17. 2)56
_(68)_

18. $7\overline{)434}$
(69)

*19. $516 \div 6$
(69)

*20. $\dfrac{279}{9}$
(69)

21. $\dfrac{267}{3}$
(69)

22. $n - 7.5 = 21.4$
(8, 45)

23. $\begin{array}{r} \$6.95 \\ \times\ \ \ \ 8 \\ \hline \end{array}$
(59)

24. $\begin{array}{r} 46 \\ \times\ 7 \\ \hline \end{array}$
(38)

25. $\begin{array}{r} 460 \\ \times\ \ \ 9 \\ \hline \end{array}$
(59)

26. $3a = 30 + 30$
(64)

27. $3^2 - 2^3$
(65)

*28. (Formulate) Write a multiplication word problem that has a product
(Inv. 1)   of 60.

29. (Conclude)   **a.** Which segment appears to be perpendicular
(17)   to segment *BC*?

   b. Name the types of angles in this triangle.

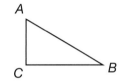

30. (Explain) During their professional baseball careers, pitcher
(16, 52)   Nolan Ryan struck out 5714 batters. Pitcher Steve Carlton struck
   out 4136 batters. How many more batters did Nolan Ryan strike out?
   Explain why your answer is reasonable.

California Mathematics Content Standards

MG 3.0, 3.3 Identify congruent figures.
MR 2.0, 2.3 Use a variety of methods, such as words, numbers, symbols, charts, graphs, tables, diagrams, and models, to explain mathematical reasoning.

• Similar and Congruent Figures

 **Power Up**

facts	Power Up I
count aloud	Count by fives from 1 to 51.
mental math	**a. Number Sense:** $21 \times 2 \times 10$
	b. Number Sense: $25 \times 2 \times 10$
	c. Number Sense: $12 \times 4 \times 10$
	d. Money: $\$5.36 + \$1.98$
	e. Measurement: Ten feet is how many inches?
	f. Estimation: Round the prices $2.58 and $6.54 to the nearest dollar and then add to estimate the total.
	g. Estimation: Round the prices $2.58 and $6.54 to the nearest 25 cents and then add to estimate the total.
	h. Calculation: $9^2 + 125 + 37$
problem solving	Choose an appropriate problem-solving strategy to solve this problem. Tazara has ten coins that total one dollar, but only one of the coins is a dime. What are the other nine coins? (There are two possibilities.)

 **New Concept**

Look at these four triangles:

Figures that are the same shape are **similar.** Figures that are the same shape and the same size are congruent.

Triangles *A* and *B* are both similar and congruent.

Triangles *B* and *C* are not congruent because they are not the same size. However, they are similar because they are the same shape. We could look at triangle *B* through a magnifying glass to make triangle *B* appear to be the same size as triangle *C*.

Triangle *A* and triangle *D* are not congruent and they are not similar. Neither one is an enlarged version of the other. Looking at either triangle through a magnifying glass cannot make it look like the other, because their sides and angles do not match.

Example 1

a. Which of these rectangles are similar?

b. Which of these rectangles are congruent?

a. Rectangles **B, C,** and **D** are similar. Rectangle *A* is not similar to the other three rectangles because it is not a "magnified" version of any of the other rectangles.

b. Rectangle **B** and **rectangle D** are congruent because they have the same shape and size.

Activity

Determining Similarity and Congruence

Material needed:
• **Lesson Activity 22**

Model Look at the shapes on the left side of **Lesson Activity 22.** Compare each shape to the figure next to it on the right, and answer each question below.

1. Is the first shape similar to the bike sign? Is the shape congruent to the bike sign? Check your answers by cutting out the shape on the left and placing it on top of the bike sign. Describe the result.

2. Is the triangle similar to the yield sign? Is the triangle congruent to the yield sign? Check your answers by cutting out the triangle and placing it on top of the yield sign. Describe the result.

3. (**Discuss**) How do you know the octagon on the left is congruent to the stop sign? Are these shapes similar?

Lesson Practice Refer to the figures below to answer problems **a** and **b.**

a. Which of these triangles appear to be similar?

b. Which of these triangles appear to be congruent?

Written Practice
Distributed and Integrated

*1. (**Analyze**) Brett can type at a rate of 25 words per minute.
(58, 59) At that rate, how many words can he type in 5 minutes? Make a table to solve this problem.

*2. Shakia has five days to read a 275-page book. If she wants to
(53, 69) read the same number of pages each day, how many pages should she read each day?

3. (**Estimate**) Umar ordered a book for $6.99, a dictionary for $8.99, and
(28, 46, 51) a set of maps for $5.99. Estimate the price for all three items. Then find the actual price.

4. Patrick practiced the harmonica for 7 weeks before his recital. How
(39) many days are equal to 7 weeks?

5. One third of the books was placed on the first shelf. What fraction of
(67) the books was not placed on the first shelf ?

*6. (**Represent**) To what decimal number is the arrow pointing? What
(43) mixed number is this?

*7. a. Multiple Choice Which two triangles appear to be congruent?
(70)

A B C D

b. **Explain** Explain your answer to part **a.**

8. Multiple Choice Cyrus ran a 5-kilometer race. Five kilometers is how
(42) many meters?

A 5 m **B** 50 m **C** 500 m **D** 5000 m

9. What is the perimeter of this triangle?
(20)

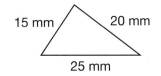

15 mm   20 mm

25 mm

*10.  **Estimate** Altogether, 117 students attend 6 different grades of a
(61, 69) small elementary school. About the same number of students attend
each grade. What is a reasonable estimate of the number of students in
each grade? Explain your answer.

*11. **Connect** The length of segment *AB* is 3.6 cm. The length of segment
(33, 45) *AC* is 11.8 cm. What is the length of segment *BC?* Write and solve a
decimal addition equation and a decimal subtraction equation.

A   B   C
●——————————●————————————————————————————————●

12. $25 − ($19.71 + 98¢)
(9, 28,
52)

13. $12 + 13 + 5 + n = 9 \times 8$
(64)

14. $5.00 − $2.92
(28)

15. $36.21 − 5.7$
(45)

16. $5 \times 6 \times 9$
(37)

17. 5×63
(38)

18. 478×6
(59)

*19. $3\overline{)147}$
(69)

*20. $7\overline{)637}$
(69)

*21. $4\overline{)136}$
(69)

22. $n + 6 = 120$
(8, 10)

23. $4w = 132$
(34, 69)

*24. $4^2 + 55$
(65)

*25. $14 + 7 \times 6$
(63)

*26. $3n = 15 + 12$
(64)

*27. $40 − 64 \div 8$
(63)

*28. **Explain** Use a formula to find the area of a square that has a side
(66) length of 4 cm. Show your work.

*29. **Estimate** Round 6.32 and 3.29 to the nearest tenth, then find
(45, 62) their sum.

30. If the diameter of a playground ball is one foot, then its radius is how
(18, 42) many inches?

*Real-World
Connection*

Constance, an international businesswoman, lives in New York,
New York. She flies to distant cities all over the world. She was
curious to know how far she traveled on some of her trips. She
did some research on the Internet and wrote down her findings
in a table.

Distance from New York (kilometers)	
Beijing, China	10,975
Paris, France	5,828
Jakarta, Indonesia	16,154
Rome, Italy	6,895

a. If Constance flies from New York to Beijing, and back again,
how far will she have traveled?

b. If she flies from New York to Jakarta, and back again, how
far will she have traveled?

c. On one business trip, Constance flew from New York to
Rome. Then she flew from Rome to Paris, and then back
to New York. If the distance from Rome to Paris is 1,120
kilometers, how far did Constance fly altogether on the
business trip?

California Mathematics Content Standards

AF 1.0, **1.5** Understand that an equation such as $y = 3x + 5$ is a prescription for determining a second number when a first number is given.

MG **2.0**, **2.1** Draw the points corresponding to linear relationships on graph paper (e.g., draw 10 points on the graph of the equation $y = 3x$ and connect them by using a straight line).

MG **2.0**, **2.2** Understand that the length of a horizontal line segment equals the difference of the x-coordinates.

MG **2.0**, **2.3** Understand that the length of a vertical line segment equals the difference of the y-coordinates.

Focus on

Coordinate Graphing

If we draw two perpendicular number lines so that they intersect at their zero points, we create an area called a **coordinate plane.** Any point within this area can be named with two numbers, one from each number line. Here we show some examples:

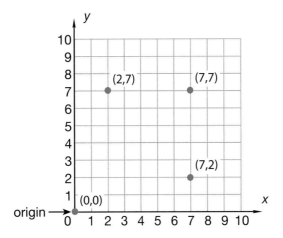

The **horizontal** number line, the line going from left to right, is called the **x-axis.** The **vertical** number line, the line going from top to bottom, is called the **y-axis.** The point where the x-axis and the y-axis intersect is called the **origin.**

The numbers in parentheses are called **coordinates,** which give a point's "address."

All coordinates are **ordered pairs** using the form (x, y).

x-coordinate, y-coordinate

(7, 2)

The first number in parentheses describes a point's horizontal distance from the origin, or the point at (0, 0). The second number in parentheses describes a point's vertical distance from the origin.

If we want to graph a point, such as (7, 2), we start at the origin. We move 7 units to the right, and then we move 2 units up. At that location we make a dot to represent the point and label it (7, 2).

1. **Model** Use **Lesson Activity 8** to practice graphing these points. Label each point with its coordinates.

   a. (5, 3)   **b.** (6, 2)   **c.** (0, 5)   **d.** (5, 0)

2. Is the point at (0, 4) on the *x*-axis or on the *y*-axis?

3. Name the coordinates of a point on the *x*-axis.

We can use letters to label points on the coordinate plane.

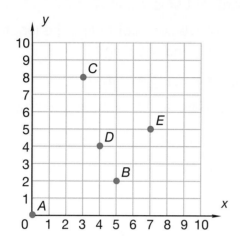

4. The coordinates of point *A* are (0, 0). What is the name for this point?

5. **Connect** Write the coordinates of each of these points.

   a. point *B*   **b.** point *C*   **c.** point *D*   **d.** point *E*

Look at points *L, M, N,* and *O.* Each point is a vertex (corner) of rectangle *LMNO.*

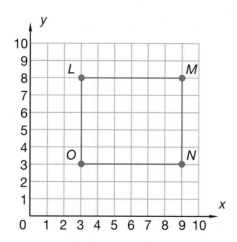

Notice that segment *LM* is one side of the rectangle. We can count the number of units between point *L* and point *M* to find the length of the rectangle. The rectangle is 6 units long.

Since the coordinates of a point are of the form (x, y), another way to find the length of segment *LM* is to subtract the x-coordinates of each endpoint.

Coordinates	(x, y)
point *M*	(9, 8)
point *L*	(3, 8)

$9 - 3 = 6$ Segment *LM* is 6 units long.

6. Conclude Which coordinates can we subtract to find the length of segment *MN*? Write the coordinates and the length.

7. Analyze What is the perimeter of rectangle *LMNO*?

8. Explain Find the perimeter of square *ABCD* without counting units. Explain your thinking.

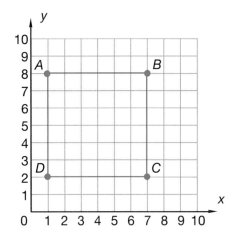

In Lesson 64, we learned how to solve an equation such as $y = 2x + 1$ when we were given the value of x. In this equation, the value of y depends on the value of x, and we can substitute different values for x to make a set of ordered (x, y) pairs. Ordered pairs are coordinates of points on the line $y = 2x + 1$ and we can use the points to graph the line.

For example, to graph the line $y = 2x + 1$, first make a table of ordered pairs by substituting different numbers for x, such as 1, 2, 3, and 4.

2x + 1 = y

x	y
1	
2	
3	
4	

The rule for this table is "multiply x by 2 and add 1."

When x is 1, y is 3.

2x + 1 = y

x	y
1	3
2	5
3	7
4	9

The result is a table of ordered pairs. The pairs are (1, 3), (2, 5), (3, 7), and (4, 9).

Now we plot the ordered pairs and connect the points. The result is a graph of the equation $y = 2x + 1$.

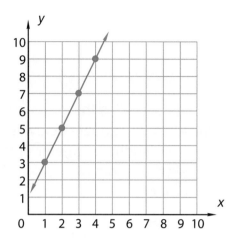

9. **Connect** Write the coordinates of another point that is on this line.

10. **Discuss** Do the coordinates $(4, 8)$ represent a point on this line? Explain why or why not.

11. **Model** Use **Lesson Activity 8** to graph five points for the equation $y = 2x + 2$.

Another type of line is a horizontal line.

This line is the graph of $y = 5$.

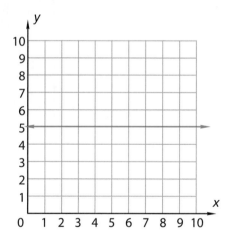

For every value of x, the value of y will always be 5. We can make a list of ordered pairs for this line.

$$y = (0 \cdot x) + 5$$

x	y
1	5
2	5
3	5

The table tells us that the points $(1, 5)$, $(2, 5)$, and $(3, 5)$ are points on the line $y = 5$.

12. (**Connect**) Write the coordinates of another point that is on this line.

13. (**Discuss**) Do the coordinates (3, 6) represent a point on this line? Explain why or why not.

14. (**Model**) Draw a graph for $y = 3$. Name the coordinates for three points on the line.

Investigate Further

a. Since a square has all four sides the same length, we know that the length of each side of the square below is 1 cm. A formula that can be used for finding the perimeter of the square is $P = 4s$ where s equals the length of one side. The perimeter of the square is 4×1 or 4 cm.

1 cm

If we use the formula $P = 4s$ to make a table of ordered pairs, then we can graph the relationship between the side length of any square and its perimeter.

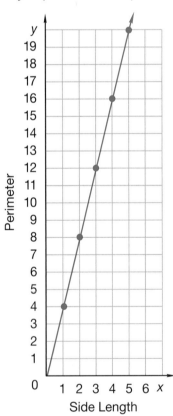

Substitute numbers for s and solve for P.

$4s = P$

s	P
1	4
2	8
3	12
4	16
5	20

(**Predict**) What point on the line represents a square that has a side length of 8 cm?

b. Use **Lesson Activity 8** to graph the equation $y = 4x$. How does the graph of the equation $y = 4x$ compare to the graph of the equation $P = 4s$?

California Mathematics Content Standards

NS **3.0**, **3.3** Solve problems involving multiplication of multidigit numbers by two-digit numbers.

NS 4.0, 4.1 Understand that many whole numbers break down in different ways (e.g., $12 = 4 \times 3 = 2 \times 6 = 2 \times 2 \times 3$).

MR 2.0, 2.6 Make precise calculations and check the validity of the results from the context of the problem.

• Multiplying by Multiples of 10

facts	Power Up H
count aloud	Count by fives from 2 to 52.
mental math	**a. Number Sense:** 300×30
	b. Number Sense: 240×10
	c. Number Sense: Counting by 5's from 5, every number Cailey says ends in 0 or 5. If she counts by 5's from 6, then every number she says ends in what digits?
	d. Fractional Parts: $\frac{1}{2}$ of 120
	e. Powers/Roots: $\sqrt{64} \div 4$
	f. Money: Cantrice bought peanuts for \$3.75 and a drink for \$2.95. What was the total cost?
	g. Estimation: Estimate the cost of 8 action figures that are each priced at \$4.95.
	h. Calculation: 9^2, $- 60$, $\div 7$, $\times 20$
problem solving	Choose an appropriate problem-solving strategy to solve this problem. Cuintan finished his 150-page book on Friday. The day before he had put the book down after reading page 120. If Cuintan read the same number of pages each day, on what day did Cuintan begin reading his book? Explain how you found your answer.

New Concept

We remember that the multiples of 10 are the numbers we say when we count by tens starting from 10. The last digit in every multiple of 10 is a zero. The first five multiples of 10 are 10, 20, 30, 40, and 50.

We may think of 20 as 2×10. So to find 34×20, we may look at the problem this way:

$$34 \times 2 \times 10$$

We multiply 34 by 2 and get 68. Then we multiply 68 by 10 and get 680.

Example 1

Write **25 × 30 as a product of 10 and two other factors. Then multiply.**

Since 30 equals 3×10, we may write 25×30 as

$$25 \times 3 \times 10$$

Three times 25 is 75, and 75 times 10 is **750.**

Analyze Is $25 \times (3 \times 10)$ the same as $25 \times (10 \times 10 \times 10)$? Why or why not?

To multiply a whole number or a decimal number by a multiple of 10, we may write the multiple of 10 so that the zero "hangs out" to the right. Below we use this method to find 34×20.

$$\begin{array}{r} 34 \\ \times\ 20 \end{array} \longleftarrow \text{zero "hangs out" to the right}$$

We first write a zero in the answer directly below the "hanging" zero.

$$\begin{array}{r} 34 \\ \times\ 20 \\ \hline 0 \end{array}$$

Then we multiply by the 2 in 20.

$$\begin{array}{r} 34 \\ \times\ 20 \\ \hline 680 \end{array}$$

Verify Is 20 the same as 10×10? Why or why not?

Example 2

To complete a spelling test, 30 students each wrote 34 different words. How many spelling words will the teacher check altogether?

We write the multiple of 10 as the bottom number and let the zero "hang out."

$$\begin{array}{r} 34 \\ \times\ 30 \end{array}$$

Next we write a zero in the answer directly below the zero in 30. Then we multiply by the 3. The teacher will check **1020 words.**

$$\begin{array}{r} 1 \\ 34 \\ \times\ 30 \\ \hline 1020 \end{array}$$

Justify How could you check the answer?

Example 3

A member of a school support staff ordered 20 three-ring binders for the school bookstore. If the cost of each binder was $1.43, what was the total cost of the order?

We write the multiple of 10 so that the zero "hangs out." We write a zero below the zero in 20, and then we multiply by the 2. We place the decimal point so that there are two digits after it. Finally, we write a dollar sign in front. The cost of the order was **$28.60.**

$$\begin{array}{r} \$1.43 \\ \times \quad 20 \\ \hline \$28.60 \end{array}$$

Lesson Practice

Multiply the factors in problems **a–f.**

a. 75×10

b. 10×32

c. $10 \times 53¢$

d. $\begin{array}{r} 26 \\ \times \quad 20 \\ \hline \end{array}$

e. $\begin{array}{r} \$1.64 \\ \times \quad 30 \\ \hline \end{array}$

f. $\begin{array}{r} 45 \\ \times \quad 50 \\ \hline \end{array}$

g. Write 12×30 as a product of 10 and two other factors. Then multiply.

Written Practice

*1. Use the information in the pictograph below to answer parts **a–c.**
(Inv. 5)

Consumed by Matt in One Day	
Water	🥛 🥛 🥛 🥛 🥛 🥛
Tea	🥛
Milk	🥛 🥛 🥛 🥛
Juice	🥛 🥛 🥛

Key: 🥛 = 1 cup

a. How many pints of liquid did Matt drink in 1 day?

b. Matt drank twice as much water as he did what other beverage?

c. About how many cups of water does Matt drink in 1 week?

2. a. What fraction of this rectangle is shaded?
(67)

b. What fraction of this rectangle is not shaded?

3. (**Estimate**) Which of these arrows could be pointing to 2500?
(Inv. 2)

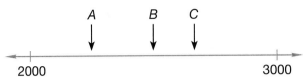

4. (**Estimate**) Zoe estimated the sum of 682 + 437 + 396 by first
(61) rounding each addend to the nearest hundred. What was Zoe's
estimate of the actual sum?

*5. (**Evaluate**) Write ordered pairs for the equation $y = 3x + 2$ when x is
(64, Inv. 7) 1, 2, 3, 4, and 5.

*6. (**Analyze**) Write a two-digit prime number that has a 3 in the ones
(56) place.

7. a. (**Estimate**) The segment below is how many centimeters long?
(42)

    b. The segment is how many millimeters long?

| cm | 1 | 2 | 3 | 4 | 5 | 6 |

8. (**Represent**) A company was sold for $7,450,000. Use words to
(47) write that amount of money.

9. If each side of a hexagon is 1 foot long, then how many inches is its
(20, 50) perimeter?

10.   93,417
(51) + 8,915

11.   42,718
(8, 52) k
    26,054

12.   1307
(51)   638
  5219
   138
 +   16

13. $100.00
(28, 52) – $ 86.32

14. 405,158
(52) – 396,370

15. 567×8
(59)

16. $30 \times 84¢$
(71)

17. $\$2.08 \times 4$
(59)

*18. 40×23
(71)

*19. 20×45
(71)

*20. 50×36
(71)

*21. $344 \div 4$
(69)

*22. $\dfrac{438}{6}$
(69)

*23. $5\overline{)355}$
(69)

24. $\sqrt{16} \times n = 100$
(Inv. 3, 34, 69)

25. $5b = 10^2$
(34, 65, 69)

*26. (**Represent**) To what decimal number is the arrow pointing? What
(43) mixed number is this?

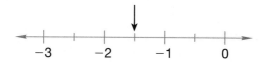

*27. a. Multiple Choice Which two rectangles appear to be congruent?
(70)

A   B   C   D

28. Find the perimeter and area of the rectangle shown
(66) at right.

5 units

3 units

29. The relationship between feet and inches is shown in the table below:
(1, 42)

Inches	12	24	36	48	60
Feet	1	2	3	4	5

a. (**Generalize**) Write a rule that describes the relationship.

b. (**Predict**) How many inches are equal to 12 feet?

*30. (Analyze) Use the coordinate graph to answer parts **a** and **b**.

(50,
Inv. 7)

a. Write the coordinates for each point.

b. If you drew a segment to connect each point, which polygon would be formed?

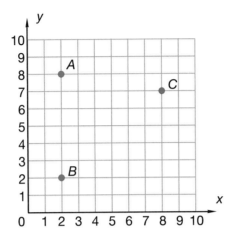

Real-World Connection

Erica and Selena recently entered *The DVD Warehouse* to buy some gifts for their friends and family. *The DVD Warehouse* includes sales tax in its prices. Refer to the tables.

Item	Cost
Comedy DVDs	$17
Action DVDs	$23
Family DVDs	$14
Anime DVDs	$11

Item	Cost
Drama DVDs	$18
History DVDs	$29
Cartoon DVDs	$12
Self-Help DVDs	$27

Erica entered the store with $40. She selected an anime DVD, a cartoon DVD, and an action DVD. Before taking them to the check-out counter she rounded the prices to the nearest 10 dollars. She estimated the cost of the 3 DVDs to be $40.

Selena entered the store with $70. She selected a drama DVD, a history DVD, and a comedy DVD. She also rounded the prices to the nearest 10 dollars before taking them to the check-out counter. She estimated the cost of the 3 DVDs to be $70.

a. When Erica brought her items to the check-out counter, the store clerk told her the total price was $46. Why was Erica's estimated price of $40 too low?

b. Selena estimated the total cost of her items to be $70. Why was $70 a safe estimate to make?

• **Division with Two-Digit Answer and a Remainder**

California Mathematics Content Standards

NS 1.0, **1.4** Decide when a rounded solution is called for and explain why such a solution may be appropriate.

NS **3.0**, **3.2** Demonstrate an understanding of, and the ability to use, standard algorithms for multiplying a multidigit number by a two-digit number and for dividing a multidigit number by a one-digit number; use relationships between them to simplify computations and to check results.

NS **3.0**, **3.4** Solve problems involving division of multidigit numbers by one-digit numbers.

MR 3.0, 3.1 Evaluate the reasonableness of the solution in the context of the original situation.

facts   Power Up H

count aloud   When we count by fives from 3, we say the numbers 3, 8, 13, 18, and so on. Count by fives from 3 to 53.

mental math

   a. Number Sense: 12×20

   b. Number Sense: 12×30

   c. Number Sense: 12×40

   d. Number Sense: $36 + 29 + 230$

   e. Money: Lucas bought a roll of film for $4.87 and batteries for $3.98. What was the total cost?

   f. Time: The baseball game started at 7:05 p.m. and lasted 1 hour 56 minutes. What time did the game end?

   g. Estimation: One mile is about 1609 meters. Round this length to the nearest hundred meters.

   h. Calculation: $\frac{1}{2}$ of 6, $\times$ 2, $\times$ 5, $-$ 16

problem solving

Levon has three colors of shirts—red, white, and blue. He has two colors of pants—black and tan. What combinations of one shirt and one pair of pants can Levon make?

Focus Strategy: Make a Diagram

(**Understand**) We are told that Levon has three colors of shirts and two colors of pants. We are asked to find the possible combinations of shirts and pants that Levon can wear.

(**Plan**) We can *make a diagram* to find all the combinations of shirt and pant colors.

Solve For each shirt, there are two colors of pants Levon can wear. We can list each shirt color and then draw two branches from each color. At the ends of the branches, we can write the color of the pants, like this:

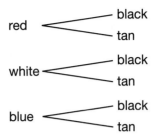

Now we list the combinations formed by the diagram. We have a total of six branches, so we find that Levon can make six different combinations of shirt and pant colors:

red, black; red, tan; white, black; white, tan; blue, black; blue, tan

Check We found six combinations that Levon can make with three different shirt colors and two different pants colors. We know our answer is reasonable because there are two combinations possible for each shirt color. There are 2 + 2 + 2, or 6 combinations for three different shirt colors.

We call the diagram we made in this problem a tree diagram, because each line we drew to connect a shirt color with a pants color is like a branch of a tree.

New Concept

The pencil-and-paper method we use for dividing has four steps: divide, multiply, subtract, and bring down. These steps are repeated until the division is complete.

Step 1: Divide.

Step 2: Multiply.

Step 3: Subtract.

Step 4: Bring down.

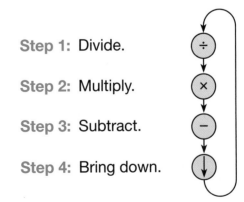

For each step we write a number. When we finish Step 4, we go back to Step 1 and repeat the steps until there are no more digits to bring down. The number left after the last subtraction is the remainder. We show the remainder in the division answer by writing it with an uppercase "R" in front.

Example 1

Divide: $5\overline{)137}$

Step 1: Divide 13 by 5 and write "2."

Step 2: Multiply 2 by 5 and write "10."

Step 3: Subtract 10 from 13 and write "3."

Step 4: Bring down 7 to make 37.

Now we repeat the same four steps:

Step 1: Divide 37 by 5 and write "7."

Step 2: Multiply 7 by 5 and write "35."

Step 3: Subtract 35 from 37 and write "2."

Step 4: There are no more digits to bring down, so we will not repeat the steps. The remainder is 2. Our answer is **27 R 2.**

$$\begin{array}{r} 2 \\ 5\overline{)137} \\ 10\!\!\downarrow \\ \hline 37 \end{array}$$

$$\begin{array}{r} 27 \\ 5\overline{)137} \\ 10 \\ \hline 37 \\ 35 \\ \hline 2 \end{array}$$

Thinking Skills

Verify

Why do we write the first digit of the quotient in the tens place?

If we divide 137 into 5 equal groups, there will be 27 in each group with 2 extra.

To check a division answer that has a remainder, we multiply the quotient (without the remainder) by the divisor and then add the remainder. For this example, we multiply 27 by 5 and then add 2.

$$\begin{array}{r} 27 \\ \times\ 5 \\ \hline 135 \end{array} \qquad \begin{array}{r} 135 \\ +\ \ 2 \\ \hline 137 \end{array} \text{ check}$$

Example 2

Three hundred seventy-five fans chartered eight buses to travel to a playoff basketball game. About how many fans were on each bus if the group was divided as evenly as possible among the eight buses?

Since we are asked to find "about how many people," we can estimate the answer. We round 375 to the nearest hundred. Instead of dividing 375 by 8, we will divide 400 by 8.

$$400 \div 8 = 50$$

There will be **about 50 people** on each bus.

Lesson Practice   Divide:

a. $3\overline{)134}$ **b.** $7\overline{)240}$ **c.** $5\overline{)88}$

d. $259 \div 8$ **e.** $95 \div 4$ **f.** $325 \div 6$

g. Shou divided 235 by 4 and got 58 R 3 for her answer. Describe how to check Shou's calculation.

h. A wildlife biologist estimates that 175 birds live in the 9-acre marsh. What is a reasonable estimate of the number of birds in each acre of the marsh? Explain why your estimate is reasonable.

Written Practice   *Distributed and Integrated*

1. There are 734 students who plan to write letters to the President. The
(37)   school has 37 boxes of envelopes. Each box contains 20 envelopes. Are there enough envelopes for everyone? How many envelopes are there together?

2. **Estimate** Clanatia went to the store with $9.12. She spent $3.92.
(28, 62)   About how much money did Clanatia have left?

3. a. Write the product of 63 using two factors.
(55, 56)

   b. Is 63 a prime number? Why or why not?

4. One fourth of the guests gathered in the living room. What fraction of
(67)   the guests did not gather in the living room?

*5. If one side of a regular triangle is 3 centimeters long, then what is its
(20, 42,   perimeter in
50)

   a. centimeters?   **b.** millimeters?

*6. **Represent** To what decimal number is the arrow pointing? What
(Inv. 2,   mixed number is this?
43)

*7. **Analyze** Moe read 30 pages a day for 14 days to finish his book. How
(39, 71)   many pages does the book have?

*8. Multiple Choice a. Which two triangles appear to be congruent?
(70)

A   B C D

b. (**Conclude**) Explain how you know.

*9. (**Explain**) Isabella estimated the product of 389 × 7 to be 2800.
(61) Explain how Isabella used rounding to make her estimate.

*10. Multiple Choice It is late afternoon. What time will it be
(13) in one hour?

   A 11:25 a.m.   **B** 5:56 a.m.

   C 4:56 p.m.   **D** 5:56 p.m.

11. (**Represent**) Compare: $\frac{3}{4}$ ◯ $\frac{4}{5}$. Draw and shade two congruent
(57) rectangles to show the comparison.

12. 4.325 − 2.5
(45)

13. 3.65 + 5.2 + 0.18
(45)

14. $50.00 − $42.60
(28, 52)

15. $17.54 + 49¢ + $15
(28, 51)

*16. 5)75
(68)

*17. 4)92
(68)

*18. 3)84
(68)

19. 398 × 6
(59)

20. 47 × 60
(71)

21. 8 × $6.25
(59)

*22. 4)136
(69)

*23. $\frac{132}{2}$
(69)

24. 6)192
(69)

25. 8n = 120
(34, 69)

26. $f \times 3^2 = 108$
(34, 65, 69)

27. 7 + 8 + 5 + 4 + n + 2 + 7 + 3 = 54
(4)

*28. Find the perimeter and area of this rectangle:
(66)

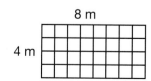

8 m

4 m

* **29.** **Explain** What is the length of segment *AB?* Explain your reasoning.
(Inv. 7)

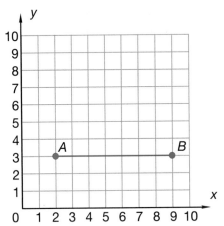

* **30.** The first four multiples of 18 are 18, 36, 54, and 72. What are the first
(55) four multiples of 180?

California Mathematics Content Standards

NS 3.0, 3.2 Demonstrate an understanding of, and the ability to use, standard algorithms for multiplying a multidigit number by a two-digit number and for dividing a multidigit number by a one-digit number; use relationships between them to simplify computations and to check results.

MR 3.0, 3.2 Note the method of deriving the solution and demonstrate a conceptual understanding of the derivation by solving similar problems.

• How the Division Algorithm Works

Look at this rectangle. It represents both $8 \times 4 = 32$ and $32 \div 4 = 8$.

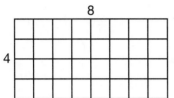

Area = 32 square units

You can think of division as finding the length of a rectangle (the quotient) when you know its area (the dividend) and its width (the divisor).

Example 1

Divide: $408 \div 6$

We can represent the steps of division by dividing a rectangle into parts. The area of the rectangle is 408 and its width is 6. The first division finds the length of a rectangle with an area of 360.

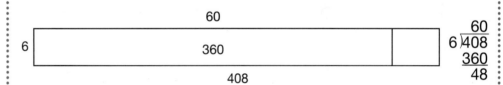

The second division finds the length of a rectangle with an area of 48.

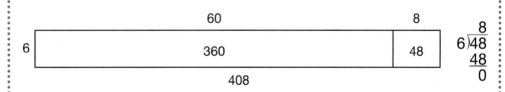

The quotient is $60 + 8$, which equals **68.**

Since multiplication and division are related operations, we may check a division answer by multiplying. $68 \times 6 = 408$

Explain Find each quotient, illustrate how to divide a rectangle representing the division, and show how to check your answer.

a. 210 ÷ 5

b. 225 ÷ 3

c. 595 ÷ 7

📐 *California Mathematics Content Standards*

NS 2.0, 2.1 Estimate and compute the sum or difference of whole numbers and positive decimals to two places.

NS 2.0, 2.2 Round two-place decimals to one decimal or the nearest whole number and judge the reasonableness of the rounded answer.

AF 1.0, 1.5 Understand that an equation such as y = 3x + 5 is a prescription for determining a second number when a first number is given.

• Capacity

facts	Power Up H
count aloud	Count down by fives from 53 to 3.
mental math	**a. Number Sense:** 21×20
	b. Number Sense: 25×30
	c. Number Sense: 25×20
	d. Number Sense: $48 + 19 + 310$
	e. Money: Julia has a gift card that is worth $50. She has used the card for $24.97 in purchases. How much value is left on the card?
	f. Time: The track meet started at 9:00 a.m. and lasted 4 hours 30 minutes. What time did the track meet end?
	g. Estimation: At sea level, sound travels about 1116 feet in one second. Round this distance to the nearest hundred feet.
	h. Calculation: $\sqrt{25}$, $\times 7$, $+ 5$, $+ 10$, $\div 10$
problem solving	Choose an appropriate problem-solving strategy to solve this problem. The charge for the taxi ride was $2.50 for the first mile and $1.50 for each additional mile. What was the charge for an 8-mile taxi ride? Explain how you solved the problem.

New Concept

Liquids such as milk, juice, paint, and gasoline are measured in the **U.S. Customary System** in **fluid ounces,** cups, pints, quarts, or gallons. The table on the next page shows the abbreviations for each of these units:

Math Language

Teaspoons and tablespoons are U.S. Customary units of measure for smaller amounts.

1 tablespoon = $\frac{1}{2}$ fluid ounce

1 teaspoon = $\frac{1}{6}$ fluid ounce

Abbreviations for U.S. Liquid Measures

fluid ounce	fl oz
cup	c
pint	pt
quart	qt
gallon	gal

The quantity of liquid a container can hold is the capacity of the container.

1 cup   1 pint   1 quart $\frac{1}{2}$ gallon   1 gallon

We can see the relationships of cups, pints, quarts, and gallons in the following diagram:

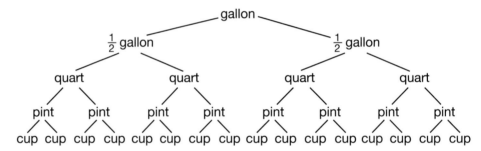

This table also shows equivalence between the units:

Equvalence Table of Units of Liquid Measures

U.S. Customary System	Metric System
8 fl oz = 1 c 2 c = 1 pt 2 pt = 1 qt 4 qt = 1 gal	1000 mL = 1 L
A **liter** is about 2 ounces more than a quart.	

Example 1

How many fluid ounces are equal to 1 pint ?

A cup is 8 fl oz and 2 cups equal 1 pint. So, a pint is equal to 2 × 8 or **16 fl oz.**

Example 2

How many pints are equal to 5 quarts?

We can find the answer by setting up a table.

Quarts	1	2	3	4	5
Pints	2	4	6	8	10

5 quarts = **10 pints**

Reading Math

We read $p = 2q$ as "the number of pints is equal to 2 times the number of quarts."

We can write a formula for converting quarts to pints. We use p for pints and q for quarts and write $p = 2q$ or $2q = p$. We can use this formula to convert any number of quarts to pints.

Generalize How could we rewrite the formula $p = 2q$ using x and y?

Example 3

Mrs. McGrath is using 4.78 L of orange juice to make punch. She poured 2.29 L of pineapple juice into the punch. To the nearest tenth of a liter, how much punch did she make?

Before adding we round each number to the nearest tenth.

4.78 L rounds to 4.8 L

2.29 L rounds to 2.3 L

Now we can add. 4.8 L + 2.3 L = **7.1 L**

Lesson Practice

a. Copy and complete this table relating gallons and quarts:

Gallons	1	2	3	4	5	6	7	8
Quarts	4	8						

b. **Predict** How many quarts is 12 gallons? Write a formula to show your answer. Use q for quarts and g for gallons.

c. One pint is 2 cups and one cup is 8 ounces. How many ounces is one pint?

d. Compare: 100 milliliters ◯ 1 liter

* 1. **Interpret** Use this circle graph to answer parts **a–d.**
(13, 19, Inv. 5)

How Franz Spent His Day

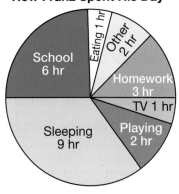

a. What is the total number of hours shown in the graph?

b. What fraction of Franz's day was spent watching TV?

c. If Franz's school day starts at 8:30 a.m., at what time does it end?

d. **Multiple Choice** Which two activities together take more than half of Franz's day?

   A sleeping and playing   **B** school and homework

   C school and sleeping   **D** school and playing

2. **Estimate** Which of these arrows could be pointing to 2250?
(Inv. 2)

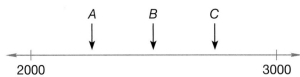

* 3. **Estimate** To find a reasonable estimate of $4.27, $5.33, and $7.64 by
(46, 61) rounding each amount to the nearest dollar before adding.

* 4. Kurt drove across the state at 90 kilometers per hour. At that rate,
(58) how far will Kurt drive in 4 hours? Make a table to solve the problem.

5. **Verify** Is the product of 3 and 7 a prime number? How do you
(56) know?

*6. a. What is the perimeter of this square?
(20, Inv. 3)

  b. If the square were to be covered with 1-inch squares, how many squares would be needed?

5 inches

7. (Evaluate) If $5 \times 10 \div 10 = 50 \div t$, what does t equal? How do you know?
(36, 63)

8. (Explain) If $(16 + 8) - f = (3 \times 8) - 7$, what does f equal? How do you know?
(9, 63)

*9. (Evaluate) If $y = 4x + 3$, then what is y when $x = 9$?
(64, Inv. 7)

10. $\$20.10$
(21, 52) $-\ \$16.45$

11. $\$98.54$
(28, 51) $+\ \$\ 9.85$

12. 380×4
(59)

13. 97×80
(71)

*14. $4\overline{)328}$
(69)

15. $\$8.63 \times 7$
(59)

16. $4.25 - 2.4$
(45)

*17. $7\overline{)375}$
(72)

*18. $5\overline{)324}$
(72)

19. $9r = 234$
(34, 69)

*20. $\dfrac{\sqrt{64}}{\sqrt{16}}$
(Inv. 3)

21. $\dfrac{287}{7}$
(69)

*22. $10 \times (6^2 + 2^3)$
(37, 65)

23. (Analyze) Find the perimeter of this rectangle
(20, 42)
  a. in centimeters.

  b. in millimeters.

1.5 cm

0.8 cm

24. The thermometer shows the outside temperature on a cold, winter day in Cedar Rapids, Iowa. What temperature does the thermometer show?
(21)

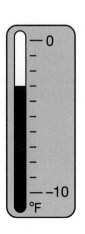

* **25.** Draw two congruent triangles and two similar rectangles.
(70)

* **26.** **Explain** How many fl oz are equal to one gallon? Explain how you
(73) found your answer.

* **27.** **Interpret** Ten students were asked to name their favorite month. Use
(Inv. 6) the results of this survey to answer questions **a** and **b**.

Five students chose July.

Four students chose August.

One student chose March.

a. What is the mode of this data?

b. What conclusion could you make based on this survey?

* **28.** **Explain** What is the length of segment *BC*? Explain your reasoning.
(Inv. 7)

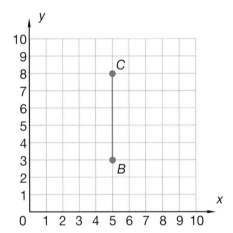

* **29.** **Estimate** Cora estimated the quotient of 261 ÷ 5 to be 50. Explain
(61, 69) how Cora used a compatible number to make her estimate.

30. **Formulate** Write and solve a subtraction word problem for the
(Inv. 1, equation $175 - t = 84$.
15)

California Mathematics Content Standards

NS 1.0, 1.7 Write the fraction represented by a drawing of parts of a figure; represent a given fraction by using drawings; and relate a fraction to a simple decimal on a number line.

MR 2.0, 2.3 Use a variety of methods, such as words, numbers, symbols, charts, graphs, tables, diagrams, and models, to explain mathematical reasoning.

MR 2.0, 2.6 Make precise calculations and check the validity of the results from the context of the problem.

• Word Problems About a Fraction of a Group

facts	Power Up H
count aloud	When we count by fives from 4, we say the numbers 4, 9, 14, 19, and so on. Count by fives from 4 to 54.
mental math	**a. Number Sense:** 25×100
	b. Number Sense: 100×40
	c. Number Sense: $12 \times 3 \times 100$
	d. Number Sense: Counting by 5's from 5, every number Raven says ends in 0 or 5. If she counts by 5's from 7, then every number she says ends in which digit?
	e. Powers/Roots: $\sqrt{4} + 3^2 + 1^2$
	f. Measurement: Abdul needs 6 quarts of water to make enough lemonade for the team. How many cups is 6 quarts?
	g. Estimation: Raoul has \$28. Does he have enough money to buy three T-shirts that cost \$8.95 each?
	h. Calculation: $\frac{1}{2}$ of 44, + 6, ÷ 7, − 4

problem solving

Choose an appropriate problem-solving strategy to solve this problem. M'Keisha solved a subtraction problem and then erased two of the digits from the problem. She gave the problem to Mae as a problem-solving exercise. Copy M'Keisha's problem on your paper, and fill in the missing digits for Mae.

$$\begin{array}{r} 123 \\ -\ 4\underline{} \\ \hline \underline{}4 \end{array}$$

New Concept

Reading Math

We can use fractions to name part of a whole, part of a group or number, and part of a distance.

We know that the fraction $\frac{1}{2}$ means that a whole has been divided into 2 parts. To find the number in $\frac{1}{2}$ of a group, we divide the total number in the group by 2. To find the number in $\frac{1}{3}$ of a group, we divide the total number in the group by 3. To find the number in $\frac{1}{4}$ of a group, we divide the total number in the group by 4, and so on.

Example 1

One half of the carrot seeds sprouted. If 84 seeds were planted, how many seeds sprouted?

We will begin by drawing a picture. The large rectangle stands for all the seeds. We are told that $\frac{1}{2}$ of the seeds sprouted, so we divide the large rectangle into 2 equal parts (into halves). Then we divide 84 by 2 and find that **42 seeds** sprouted.

84 seeds

$\frac{1}{2}$ sprouted. $\left\{ \boxed{42 \text{ seeds}} \right.$

$\frac{1}{2}$ did not sprout. $\left\{ \boxed{42 \text{ seeds}} \right.$

$$2\overline{)84 \text{ seeds}}^{\,42 \text{ seeds}}$$

Discuss How can we use addition to check the answer?

Example 2

On Friday, one third of the 27 students purchased lunch in the school cafeteria. How many students purchased lunch on Friday?

We start with a picture. The whole rectangle stands for all the students. Since $\frac{1}{3}$ of the students purchased lunch, we divide the rectangle into 3 equal parts. To find how many students are in each part, we divide 27 by 3 and find that **9 students** purchased a lunch on Friday.

27 students

$\frac{1}{3}$ purchased lunch. $\left\{ \boxed{9 \text{ students}} \right.$

$\frac{2}{3}$ did not purchase lunch. $\left\{ \boxed{\begin{array}{c} 9 \text{ students} \\ 9 \text{ students} \end{array}} \right.$

$$3\overline{)27 \text{ students}}^{\,9 \text{ students}}$$

Justify Explain why the answer is correct.

Example 3

One fourth of the team's 32 points were scored by Thi. How many points did Thi score?

We draw a rectangle. The whole rectangle stands for all 32 points. Thi scored $\frac{1}{4}$ of the points, so we divide the rectangle into 4 equal parts. We divide 32 by 4 and find that each part is 8 points. Thi scored **8 points.**

$\frac{1}{4}$ scored by Thi

$\frac{3}{4}$ not scored by Thi

32 points

| 8 points |
| 8 points |
| 8 points |
| 8 points |

$$\frac{8 \text{ points}}{4)\overline{32 \text{ points}}}$$

Justify Explain why the answer is correct.

Example 4

What is $\frac{1}{5}$ of 40?

We draw a rectangle to stand for 40. We divide the rectangle into five equal parts, and we divide 40 by 5. Each part is 8, so $\frac{1}{5}$ of 40 is **8.**

$\frac{1}{5}$ of 40

$\frac{4}{5}$ of 40

40

| 8 |
| 8 |
| 8 |
| 8 |
| 8 |

$$\frac{8}{5)\overline{40}}$$

Lesson Practice   Draw a picture to solve each problem:

a. What is $\frac{1}{3}$ of 60?

b. What is $\frac{1}{2}$ of 60?

c. What is $\frac{1}{4}$ of 60?

d. What is $\frac{1}{5}$ of 60?

e. One half of the 32 children were boys. How many boys were there?

f. One third of the 24 coins were quarters. How many quarters were there?

1. There are 77 students who plan to visit a small history museum for their
(37, 71)   class. The museum allows groups of 5 students to enter the museum
every 10 minutes. How many minutes would it take for all 77 students
to enter the museum? How many groups would be let in altogether?

2. (**Analyze**) Monty ran the race 12 seconds faster than Ivan. Monty ran
(16)   the race in 58 seconds. Ivan ran the race in how many seconds?

3. (**Analyze**) There were 4 rooms. One fourth of the 56 guests were
(74)   gathered in each room. How many guests were in each room?

4. (**Analyze**) How many 6-inch-long sticks can be cut from a 72-inch-long
(53)   stick?

5. **Multiple Choice** One fifth of the leaves had fallen. What fraction of
(67)   the leaves had *not* fallen?

A $\frac{2}{5}$ **B** $\frac{3}{5}$ **C** $\frac{4}{5}$ **D** $\frac{5}{5}$

6. (**Estimate**) Which of these arrows could be pointing to 5263?
(Inv. 2)

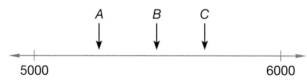

*7. (**Explain**) If $y = 5x + 2$, then what is y when $x = 20$? Explain your
(64)   thinking.

*8. (**Conclude**) Which word makes the following sentence false?
(50, 70)
All squares are _____

A polygons **B** rectangles **C** similar **D** congruent

*9. (**Explain**) Cleon would like to estimate the difference between
(22) $579 and $85. Explain how Cleon could use rounding to make an
estimate.

*10. The triangle at right is equilateral.
(20, 42, 50)

   a. How many millimeters is the perimeter of the triangle?

   b.  Describe the angles.

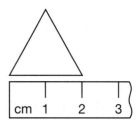

11. Three liters equals how many milliliters?
(73)

*12. Wilma runs 5 miles in 1 hour. At that rate, how long would it take her to run 40 miles? Make a table to solve the problem.
(58)

13. $2n = 150$
(34, 69)

14. $24.25 - (6.2 + 4.8)$
(9, 45)

15. $\begin{array}{r} 103{,}279 \\ +\ \ 97{,}814 \\ \hline \end{array}$
(51)

16. $\begin{array}{r} \$36.14 \\ +\ \$27.95 \\ \hline \end{array}$
(28, 50)

17. $\begin{array}{r} 39{,}420 \\ -\ 29{,}516 \\ \hline \end{array}$
(52)

18. $\begin{array}{r} \$60.50 \\ -\ \ \ \ \ \ n \\ \hline \$43.20 \end{array}$
(8, 28, 52)

19. $\begin{array}{r} 604 \\ \times\ \ \ \ 9 \\ \hline \end{array}$
(59)

20. $\begin{array}{r} 87 \\ \times\ \ 60 \\ \hline \end{array}$
(71)

21. $\begin{array}{r} \$6.75 \\ \times\ \ \ \ \ 4 \\ \hline \end{array}$
(59)

*22. $7\overline{)243}$
(72)

*23. $5\overline{)323}$
(72)

*24. $\begin{array}{r} n \\ +\ 1467 \\ \hline 2459 \end{array}$
(8, 52)

*25. $7\overline{)429}$
(72)

26. $189 \div 6$
(72)

27. $472 \div 8$
(69)

28. $9w = 9^2 + (9 \times 2)$
(34, 64, 65)

*29. $3\overline{)288}$
(69)

*30. **Conclude** Use the coordinate graph to answer parts **a** and **b**.

(Inv. 7)

   a. Name the coordinates of each point.

   b. Assume points *D, E,* and *F* are each a vertex of a square. Point *G* will be placed at the fourth vertex. Name the coordinates of point *G*.

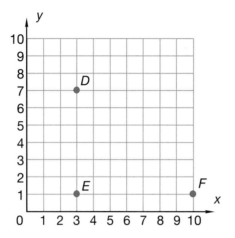

Real-World Connection

Every Saturday morning, Mahina stops by Ms. Hsiao's fruit stand and buys 1 pummelo and 1 carambola. Ms. Hsiao charges her $1.36 for a pummelo and $0.72 for a carambola, not including tax. Next week, Ms. Hsiao is going to round the prices of her fruit to the nearest dime. She believes this will make giving change easier for her customers.

   a. If Ms. Hsiao rounds the prices of her fruit to the nearest dime, how much will Mahina pay for a pummelo and a carambola?

   b. By rounding her prices to the nearest dime, how much more money will Mahina have to pay?

   c. By rounding the prices, the cost of a pummelo increased but the price of a carambola decreased. Explain why Mahina still had to pay more money.

California Mathematics Content Standards

NS 1.0, **1.4** Decide when a rounded solution is called for and explain why such a solution may be appropriate.

NS **3.0**, **3.2** Demonstrate an understanding of, and the ability to use, standard algorithms for multiplying a multidigit number by a two-digit number and for dividing a multidigit number by a one-digit number; use relationships between them to simplify computations and to check results.

NS **3.0**, **3.4** Solve problems involving division of multidigit numbers by one-digit numbers.

• Division Answers Ending with Zeros

facts	Power Up H
count aloud	Count down by fives from 54 to 4.
mental math	The sum of 38 and 17 is 55. If we make 38 larger by 2 and 17 smaller by 2, then the addition is 40 + 15. The sum is still 55, but the mental addition is easier. Before finding the following sums, make one number larger and the other smaller so that one of the numbers ends in zero.

    a. Number Sense: 38 + 27

    b. Number Sense: 48 + 24

    c. Number Sense: 59 + 32

    d. Number Sense: 57 + 26

    e. Money: $6.49 + $2.99

    f. Measurement: How many cups is one pint?

    g. Estimation: Choose the more reasonable estimate for the temperature inside a refrigerator: 3°C or 30°C.

    h. Calculation: 2 × 9, + 29, + 53, ÷ 10

problem solving	Choose an appropriate problem-solving strategy to solve this problem. Sid wants to know the distance around the trunk of the big oak tree at the park. He knows the distance around the trunk is more than one yard. Sid has some string and a yardstick. How can he measure the distance around the trunk of the tree in inches?

Sometimes division answers end with a zero. It is important to continue the division until all the digits inside the division box have been used. Look at this problem:

Two hundred pennies are separated into 4 equal piles. How many pennies are in each pile?

Thinking Skills

Verify

Why do we write the first digit of the quotient in the tens place?

This problem can be answered by dividing 200 by 4. First we divide 20 by 4. We write a 5 in the quotient. Then we multiply and subtract.

$$\begin{array}{r} 5 \\ 4\overline{)200} \\ \underline{20} \\ 0 \end{array}$$

The division might look complete, but it is not. The answer is not "five pennies in each pile." That would total only 20 pennies. There is another zero inside the division box to bring down. So we bring down the zero and divide again. Zero divided by 4 is 0. We write 0 in the quotient, multiply, and then subtract. The quotient is **50.**

$$\begin{array}{r} 50 \\ 4\overline{)200} \\ \underline{20\downarrow} \\ 00 \\ \underline{0} \\ 0 \end{array}$$

Check:

$$\begin{array}{r} 50 \\ \times\ 4 \\ \hline 200 \end{array}$$

We check our work by multiplying the quotient, 50, by the divisor, 4. The product should equal the dividend, 200. The answer checks. We find that there are 50 pennies in each pile.

Sometimes there will be a remainder with a division answer that ends in zero. We show this in the following example.

Example 1

Thinking Skills

Verify

Why do we write the first digit of the quotient in the tens place?

Divide: $3\overline{)121}$

We begin by finding $3\overline{)12}$. Since 12 divided by 3 is 4, we write "4" above the 2. We multiply and subtract, getting 0, but we are not finished. We bring down the last digit of the dividend, which is 1.

$$\begin{array}{r} 4 \\ 3\overline{)121} \\ \underline{12} \\ 0 \end{array}$$

Now we divide 01 (which means 1) by 3. Since we cannot make an equal group of 3 if we have only 1, we write "0" in the ones place of the quotient. We then multiply zero by 3 and subtract. The remainder is 1.

$$\begin{array}{r} 40\ R\ 1 \\ 3\overline{)121} \\ \underline{12} \\ 01 \\ \underline{0} \\ 1 \end{array}$$

Example 2

Mr. Griffith drove 254 miles in 5 hours. About how many miles did he drive each hour?

To find "about how many miles" Mr. Griffith drove each hour, we can use compatible numbers to estimate. Since 250 is close to 254 and is divisible by 5, we divide 250 by 5 to estimate.

$$250 \text{ miles} \div 5 \text{ hours} = 50 \text{ miles each hour}$$

Mr. Griffith drove **about 50 miles** each hour.

Lesson Practice

Divide:

a. $3\overline{)120}$ **b.** $4\overline{)240}$ **c.** $5\overline{)152}$

d. $4\overline{)121}$ **e.** $3\overline{)91}$ **f.** $2\overline{)41}$

g. (Estimate) The employees in the shipping department of a company loaded 538 boxes into a total of 6 railcars. They put about the same number of boxes into each railcar. About how many boxes are in each railcar? Explain how you found your answer.

Written Practice

Distributed and Integrated

1. Cecilia skated 27 times around the rink forward and 33 times around the rink backwards. In all, how many times did she skate around the rink?
(RF1, RF5)

2. Nectarines cost 68¢ per pound. What is the price for 3 pounds of nectarines?
(39)

* 3. (Analyze) In bowling, the sum of Amber's score and Bianca's score was equal to Consuela's score. If Consuela's score was 113 and Bianca's score was 55, what was Amber's score? Write an equation to show your work.
(12, 14)

* 4. One third of the 84 students were assigned to each room. How many
(68) students were assigned to each room? Draw a picture to explain how
you found your answer.

5. Round 2250 to the nearest thousand.
(40)

6. **Represent** Use digits to write the number five million, three hundred
(47, 48) sixty-five thousand in standard form. Then round it to the nearest
hundred thousand.

* 7. **Classify** Write the name of each polygon:
(50, 70)
   a. five sides

   b. ten sides

   c. six sides

   d. eight sides

8. **Estimate** The tip of this shoelace is how many millimeters long?
(42)

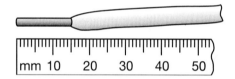

* 9. **Conclude** Choose the more reasonable measure for **a** and **b.**
(73)
   a. milk for a bowl of cereal: 2 pt or 4 oz

   b. a full pail of water: 1 pt or 1 gal

10. According to this calendar, what is the date of the last
(RF12) Tuesday in February 2019?

FEBRUARY 2019						
S	M	T	W	T	F	S
					1	2
3	4	5	6	7	8	9
10	11	12	13	14	15	16
17	18	19	20	21	22	23
24	25	26	27	28		

11. **Represent** Forty-two thousand, seven hundred is how much greater
(27, 52) than thirty-four thousand, nine hundred?

12. Find the perimeter and area of this rectangle:
(66)

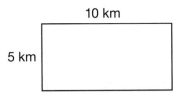

10 km

5 km

*

13. (**Conclude**) Use the coordinate graph to answer parts **a** and **b**.
(Inv. 7)
    a. Name the coordinates for point *P* and point *Q*.

    b. What is the length of segment *PQ*?

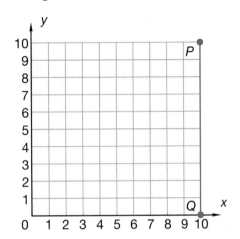

14. 6743 − (507 × 6)
(52, 59, 63)

15. $70.00 − $63.17
(28, 52)

16. 3 × 7 × 0
(24)

17. $8.15 × 6
(59)

18. 67¢ × 20
(71)

19. 4.5 + 0.52 + 1.39
(45)

*

20. $5\overline{)323}$
(72)

*

21. $4\overline{)159}$
(72)

*

22. 329 ÷ 6
(72)

*

23. $\dfrac{180}{3}$
(75)

*

24. $5^3 ÷ 5$
(65, 68)

*

25. 241 ÷ 8
(75)

26. 4*n* = 200
(34, 75)

27. 7*d* = 105
(34, 69)

28.
(8, 10, 11)
    473
    286
  + *n*
  ―――
    943

29. 1 + 12 + 3 + 14 + 5 + 26
(11)

* **30.** The bar graph shows the average lifespan in years of several animals.
$_{(19,}^{(19,}$
$_{Inv.\ 5)}$ Use the graph to solve the problems that follow.

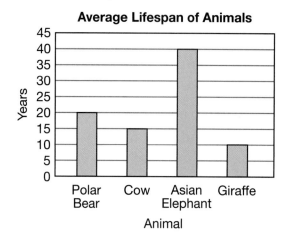

Average Lifespan of Animals

a. Write the names of the animals in order from longest to shortest average lifespan.

b. What fraction of the average lifespan of an Asian elephant is the average lifespan of a polar bear?

c. When compared to the average lifespan of a giraffe, how many times greater is the average lifespan of an Asian elephant?

- **Finding Information to Solve Problems**

California Mathematics Content Standards

NS 1.0, **1.8** Use concepts of negative numbers (e.g., on a number line, in counting, in temperature, in "owing").

SDAP 1.0, 1.3 Interpret one-and two-variable data graphs to answer questions about a situation.

MR 1.0, 1.2 Determine when and how to break a problem into simpler parts.

MR 2.0, 2.2 Apply strategies and results from simpler problems to more complex problems.

Power Up

facts	Power Up G
count aloud	Count by fives from 1 to 51.
mental math	Before adding, make one number larger and the other number smaller.

a. Number Sense: 49 + 35

b. Number Sense: 57 + 35

c. Number Sense: 28 + 44

d. Number Sense: 400 × 30

e. Money: KaNiyah owes her brother $10.00. She only has $4.98. How much more money does she need to repay her brother?

f. Measurement: Seven feet is 84 inches. A dolphin that is 7 feet 7 inches long is how many inches long?

g. Estimation: Each half-gallon of milk costs $2.47. Round this price to the nearest 25 cents and estimate the cost of 3 half-gallon containers of milk.

h. Calculation: $\sqrt{25}$, × 2, ÷ 5, × 15, + 48

problem solving

Choose an appropriate problem-solving strategy to solve this problem. Phuong and LaDonna are planning a hike in the park. The map of the park's trails is shown at right.

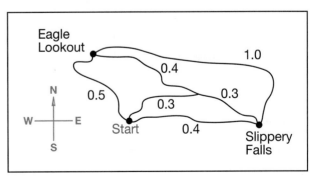

Phuong and LaDonna will start at the point labeled "Start."

They want to visit both Eagle Lookout and Slippery Falls. What is the shortest distance Phuong and LaDonna can hike in order to visit both points and then return to where they started? All distances shown are in kilometers.

New Concept

Part of the problem-solving process is finding the information needed to solve a problem. Sometimes we need to find information in graphs, tables, pictures, or other places. In some cases, we might be given more information than we need to solve a problem. In this lesson we will be finding the information we need to solve a problem.

Example 1

Read this information. Then answer the questions that follow.

The school elections were held on Friday, February 2. Tejana, Lily, and Taariq ran for president. Lily received 146 votes, and Tejana received 117 votes. Taariq received 35 more votes than Tejana.

a. How many votes did Taariq receive?

b. Who received the most votes?

c. Speeches were given on the Tuesday before the elections. What was the date on which the speeches were given?

a. Taariq received 35 more votes than Tejana, and Tejana received 117 votes. So we add 35 to 117 and find that Taariq received **152 votes.**

b. Taariq received the most votes.

c. The elections were on Friday, February 2. The Tuesday when the speeches were presented was 3 days before that. We count back 3 days: February 1, January 31, January 30. The speeches were given on Tuesday, **January 30.**

Example 2

The balance in Emily's bank account is $50. If she writes a check for $70, what number represents the balance in her account?

The amount of money in a bank account is called a balance. When Emily writes a check, the balance decreases. When she makes a deposit, the balance increases.

We can use a number line to illustrate the balance of Emily's account. We start at zero and move to the right from 0 to 50 to represent $50.

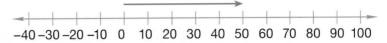

Writing a check is the same as taking dollars away. We can represent writing a check of $70 by moving to the left. We start at 50 and count by tens to 70. We can see that we end at –20.

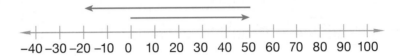

We can represent the balance in Emily's account with a negative number, **–$20.** This means that Emily owes the bank $20.

Discuss How can we write an equation to represent this problem?

Example 3

Use the double bar graph to solve these problems.

a. Did the population of Fairview increase or decrease from 1990 to 2004? Write an equation to show the population change.

b. Which two cities had the same population change from 1990 to 2004?

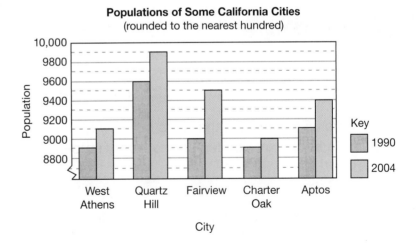

When we need to solve a problem using information in a graph, it is the same as a problem with too much information. We need to look at the graph and find the information we need.

a. We can see that Fairview had a population of 9000 in 1990. It increased to 9500 by 2004. We can show this with the equation $9000 + n = 9500$. We count on or subtract to find that $n = 500$.

The population of Fairview increased by **500** people between 1990 and 2004.

b. We compare each pair of bars to find an increase that is the same. We can see that **Quartz Hill and Aptos** each increased by 300 people between 1990 and 2004.

Lesson Practice Read this information. Then solve the problems that follow.

Terell did yard work on Saturday. He worked for 3 hours in the morning and 4 hours in the afternoon. He was paid $6 for every hour he worked.

a. How many hours did Terell work in all?

b. How much money did Terell earn in the morning?

c. How much money did Terell earn in all?

d. (Analyze) Daniel had $30 in the bank. He wrote a check for $45. What number represents the balance of his account?

e. (Interpret) Use the double bar graph in Example 3 to answer this question. Which city increased by 100 people between 1990 and 2004?

Written Practice *Distributed and Integrated*

(Formulate) Write and solve equations for problems **1–4.**

1. There were 35 students in the class but only 28 math books. How many
(16) more math books are needed so that every student in the class has a math book?

2. Each of the 7 children slid down the water slide 11 times. How many
(39) times did they slide in all?

*3. A bowling lane is 60 feet long. How many yards is 60 feet?
(42, 53)

4. Wei carried the baton four hundred forty yards. Eric carried it eight
$_{(3, 27, 51)}$ hundred eighty yards. Joe carried it one thousand, three hundred twenty yards, and Bernardo carried it one thousand, seven hundred sixty yards. In all, how many yards was the baton carried?

5. One third of the members voted "no." What fraction of the members
$_{(67)}$ did not vote "no"?

*6. ✎ **Explain** Marissa would like to estimate the sum of 6821 + 4963.
$_{(40, 61)}$ Explain how Marissa could use rounding to make an estimate.

*7. **Generalize** Use this table to write a formula that can be used to
$_{(73)}$ convert any number of pints to ounces. Use *o* for ounces and *p* for pints.

Pints	1	2	3	4	5
Ounces	16	32	48	64	80

8. **Represent** Write the number that is 500,000 greater than 4,250,000.
$_{(47, 51)}$

*9. **Interpret** Use the line graph below to answer parts **a–c.**
$_{(Inv. 5)}$

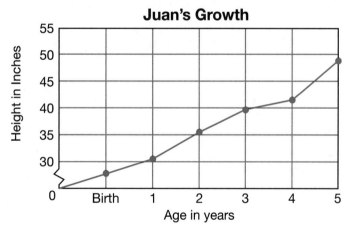

a. About how many inches tall was Juan on his second birthday?

b. About how many inches did Juan grow between his third and fifth birthdays?

c. Copy and complete this table using information from the line graph:

Juan's Growth

Age	Height
At birth	25 inches
1 year	
2 years	

*10. **Represent** On the last Friday in May, one fourth of the 280 students in a school were away on a field trip. How many students were on the field trip? Draw a picture to solve the problem.
(74)

*11. **Analyze** Simplify each expression. Remember to use the order of operations.
(63)

a. $5 \times 3 + 35 \div 7$

b. $81 \div 9 \times 2 + 7$

c. $100 - (8 + 3) \times 6$

d. $5 \times (10 - 6) \div 5$

*12. The table shows the number of vacation days Carson earns at work:
(Inv. 7, 76)

Days Worked	Vacation Days Earned
30	1
60	2
90	3
120	4
150	5
180	6

a. **Generalize** Write a word sentence that describes the relationship of the data.

b. **Predict** Use the word sentence you wrote to predict the number of days Carson needs to work to earn 10 vacation days.

13.
(28, 51)

$60.75
+ $95.75

14.
(28, 52)

$16.00
− $15.43

15.
(45)

3.15
− 3.12

16.
(71)

32
× 30

17.
(59)

465
× 7

18.
(59)

$0.98
× 6

19. 425 ÷ 6
(75)

*20. 462 ÷ 9
(72)

*21. 159 ÷ 4
(72)

22. $3r = 150$
(34, 75)

23. $10^2 + t = 150$
(63, 65)

24. $1 + 7 + 2 + 6 + 9 + 4 + n = 37$
(4)

25. a. If the 3-inch square is covered with 1-inch squares, how many of the 1-inch squares are needed?

b. What is the area of the 3-inch square?
(Inv. 3)

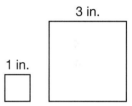

26. a. What is the perimeter of this triangle?
(17, 20, 45)

b. Describe the angles of the triangle.

*27. (Evaluate) If $y = 3x + 7$, then what is y when $x = 4$?
(64, Inv. 7)

*28. (Conclude) Is $\overline{AB} \perp \overline{BC}$? Why or why not?
(17)

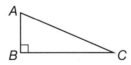

29. Multiple Choice Three of these triangles are congruent. Which triangle is *not* one of the three congruent triangles?
(70)

A   **B**   **C**   **D**

* 30. The radius of this circle is 1.2 cm. What is the diameter of the
(18, 45) circle?

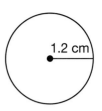

1.2 cm

Real-World Connection

There are four people in the Escobar family and five in the Greene family.

The Escobar family and the Greene family went on a vacation together. They rented a beach house for 3 days. The cost of renting the beach house for 3 days was $783.

    a. What was the cost per person of renting the beach house?

    b. If the two families split the cost, how much should the Escobars pay given that they have 4 members in their family?

    c. How much should the Greenes pay given that they have 5 people in their family?

✎ *California Mathematics Content Standards*

NS 1.0, 1.5 Explain different interpretations of fractions, for example, parts of a whole, parts of a set, and division of whole numbers by whole numbers; explain equivalents of fractions (see Standard 4.0).

NS 1.0, 1.7 Write the fraction represented by a drawing of parts of a figure; represent a given fraction by using drawings; and relate a fraction to a simple decimal on a number line.

MR 2.0, 2.6 Make precise calculations and check the validity of the results from the context of the problem.

• Fraction of a Set

facts	Power Up G
count aloud	Count by fives from 2 to 52.
mental math	Before adding, make one number larger and the other number smaller in problems **a–c.**

    a. Number Sense: $55 + 47$

    b. Number Sense: $24 + 48$

    c. Number Sense: $458 + 33$

    d. Number Sense: 15×30

    e. Money: Renee bought a pair of gloves for $14.50 and a hat for $8.99. What was the total cost of the items?

    f. Measurement: Compare: 2 miles ◯ 10,000 feet

    g. Estimation: An *acre* is a measurement of land. A square plot of land that is 209 feet on each side is about 1 acre. Round 209 feet to the nearest hundred feet.

    h. Calculation: 7^2, $- 1$, $\div 8$, $+ 4$, $- 4$, $\div 6$

problem solving	Choose an appropriate problem-solving strategy to solve this problem. Colby wants to cover his bulletin board with square sheets of paper that are 1 foot on each side. His bulletin board is 5 feet wide and 3 feet tall. If Colby has already cut 12 squares of paper, how many more squares does he need to cut? Explain how you found your answer.

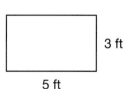

3 ft

5 ft

New Concept

Thinking Skills

Discuss

How can we check the answer?

There are seven circles in the set below. Three of the circles are shaded. The fraction of the set that is shaded is $\frac{3}{7}$.

$\frac{3}{7}$ Three circles are shaded.
There are seven circles in all.

The total number of members in the set is the denominator (bottom number) of the fraction. The number of members named is the numerator (top number) of the fraction.

Example 1

What fraction of the triangles is not shaded?

Thinking Skills

Verify

How can we check the answer?

The denominator of the fraction is 9, because there are 9 triangles in all.
The numerator is 5, because 5 of the 9 triangles are not shaded.
So the fraction of triangles that are not shaded is $\frac{5}{9}$.

Example 2

In a class of 25 students, there are 12 girls and 13 boys. What fraction of the class is girls?

Twelve of the 25 students in the class are girls. So the fraction of the class that is girls is $\frac{12}{25}$.

Lesson Practice

a. What fraction of the set is shaded?

b. What fraction of the set is not shaded?

c. In a class of 27 students, there are 14 girls and 13 boys. What fraction of the class is boys?

d. In the word ALABAMA, what fraction of the letters are A's?

Written Practice *Distributed and Integrated*

1. **Multiple Choice** To prepare for a move to a new building, the
(71) employees of a library spent an entire week packing books in boxes.
On Monday, the employees packed 30 books in each of 32 boxes. How
many books did those boxes contain?

A 9600 books   **B** 960 books   **C** 320 books   **D** 350 books

2. The movie was 3 hours long. If it started at 11:10 a.m., at what time did
(13) it end?

3. **Explain** Jonathan is reading a 212-page book. If he has finished
(15, 22) 135 pages, how many pages does he still have to read? Explain why
your answer is reasonable.

4. Khalil, Julian, and Elijah each scored one third of the team's 42 points.
(74) They each scored how many points?

5. **Estimate** A family has $4182 in a savings account. Round the
(40) number of dollars in the account to the nearest thousand.

*6. **Explain** The shirt was priced at $16.98. The tax was $1.02. Sam
(28, 51, 52) paid the clerk $20. How much money should Sam get back? Explain
your thinking.

*7. What fraction of the letters in the following word are I's?
(77)   S U P E R C A L I F R A G I L I S T I C E X P I A L I D O C I O U S

*8. **Analyze** Use the information below to answer parts **a–c.**
(76)
*In the first 8 games of this season, the Rio Hondo football team
won 6 games and lost 2 games. They won their next game by a
score of 24 to 20. The team will play 12 games in all.*

a. In the first nine games of the season, how many games did Rio
Hondo win?

b. Rio Hondo won its ninth game by how many points?

c. What is the greatest number of games Rio Hondo could win this
season?

9. Compare: $3 \times 4 \times 5 \bigcirc 5 \times 4 \times 3$
(23, 26)

10. $m - 137 = 257$
(6, 8)

11. $n + 137 = 257$
(8)

12. $1.45 + 2.4 + 0.56 + 7.6$
(45)

13. $5.75 - (3.12 + 0.5)$
(9, 45)

14.
(71)
$$\begin{array}{r} 38 \\ \times\ 50 \\ \hline \end{array}$$

15.
(59)
$$\begin{array}{r} 472 \\ \times\ 9 \\ \hline \end{array}$$

16.
(59)
$$\begin{array}{r} \$6.09 \\ \times\ 6 \\ \hline \end{array}$$

* 17. $9\overline{)892}$
(72)

* 18. $4\overline{)286}$
(72)

* 19. $3\overline{)109}$
(72)

20. $121 \div 3$
(75)

21. $122 \div 4$
(75)

* 22. $7\overline{)566}$
(75)

23. $9^2 = 9n$
(34, 65)

24. $5w = 5 \times 10^2$
(34, 64, 65)

* 25. a. (**Model**) Use a ruler to find the perimeter of the
(20, 42) rectangle at right in millimeters.

b. (**Analyze**) Draw a rectangle that is similar to the rectangle in part **a** and whose sides are twice as long. What is the perimeter in centimeters of the rectangle you drew?

*26. (**Evaluate**) Simplify each expression when $n = 8$.
(37, 63)
    a. $82 - 88 \div n$ **b.** $6 \times (22 + n) - 5$

*27. (**Evaluate**) If $y = 2x + 9$, then what is y when $x = 1, 2,$ and 3?
(64, Inv. 7)

*28. (**Represent**) Write three ordered pairs for the equation in problem **27**.
(Inv. 7) Then use **Lesson Activity 8** to graph the ordered pairs.

* 29. (**Generalize**) Use this table to write a formula that can be used to
(73) convert any number of quarts to cups. Use q for quarts and c for cups.

Quarts	1	2	3	4	5
Cups	4	8	12	16	20

*30. (**Analyze**) What decimal number names the point marked B on this
(43) number line?

◢ *California Mathematics Content Standards*
MG 3.0, 3.3 Identify congruent figures.
MG 3.0, 3.5 Know the definitions of a right angle, an acute angle, and an obtuse angle. Understand that 90°, 180°, 270°, and 360° are associated, respectively, with $\frac{1}{4}$, $\frac{1}{2}$, $\frac{3}{4}$, and full turns.
MR 3.0, 3.2 Note the method of deriving the solution and demonstrate a conceptual understanding of the derivation by solving similar problems.

• Measuring Turns

facts	Power Up G
count aloud	Count by fives from 3 to 53.
mental math	**a. Number Sense:** 35×100

b. Number Sense: Counting by 5's from 5, every number Ramon says ends in 0 or 5. If he counts by 5's from 8, then every number he says ends in which digit?

c. Fractional Parts: $\frac{1}{2}$ of $31.00

d. Measurement: Jenna jogged 3 kilometers. How many meters is that?

e. Money: The box of cereal cost $4.36. Tiana paid with a $5 bill. How much change should she receive?

f. Time: Rodrigo's school day lasts 7 hours. If Rodrigo attends school Monday through Friday, how many hours is he at school each week?

g. Estimation: Each CD costs $11.97. Estimate the cost of 4 CDs.

h. Calculation: $\frac{1}{2}$ of 88, $+$ 11, $\div$ 11

problem solving

Choose an appropriate problem-solving strategy to solve this problem. Robby is mailing an envelope that weighs 6 ounces. The postage rates are 39¢ for the first ounce and 24¢ for each additional ounce. If Robby pays the postal clerk $2.00 for postage, how much money should he get back?

As Micah rides a skateboard, we can measure his movements. We might use feet or meters to measure the distance Micah travels. To measure Micah's turns, we may use **degrees.** Just as for temperature measurements, we use the degree symbol (°) to stand for degrees.

If Micah makes a **full turn,** then he has turned 360°. If Micah makes a **half turn,** he has turned 180°. A **quarter turn** is 90°.

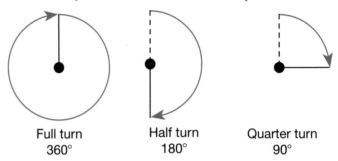

| Full turn | Half turn | Quarter turn |
| 360° | 180° | 90° |

Besides measuring the amount of turn, we may also describe the direction of a turn as **clockwise** or **counterclockwise.**

Clockwise turn   Counterclockwise turn

For instance, we tighten a screw by turning it clockwise, and we loosen a screw by turning it counterclockwise.

Rotations and Degrees

Stand and perform these activities as a class.

(**Model**) Face the front of the room and make a quarter turn to the right.

Discuss How many degrees did you turn? Did you turn clockwise or counterclockwise?

Return to your original position by turning a quarter turn to the left.

a. How many degrees did you turn? Did you turn clockwise or counterclockwise?

Face the front of the room and make a half turn either to the right or to the left.

b. How many degrees did you turn? Is everyone facing the same direction?

Start by facing the front. Then make a three quarter turn clockwise.

c. How many degrees did you turn? How many more degrees do you need to turn clockwise in order to face the front?

Example 1

Mariya and Irina were both facing north. Mariya turned 90° clockwise and Irina turned 270° counterclockwise. After turning, in which directions were the girls facing?

Below we show the turns Mariya and Irina made.

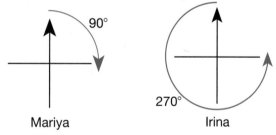

After turning 90° clockwise, Mariya was facing east. After turning 270° counterclockwise, Irina was also facing east. (Each quarter turn is 90°, so 270° is three quarters of a full turn.) Both girls were facing **east** after their turns.

Example 2

Describe the amount and the direction of a turn about point *A* that would position △II on △I.

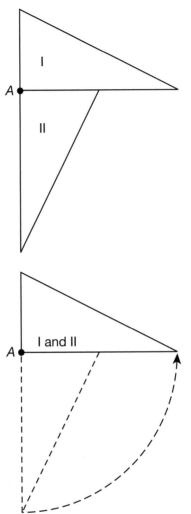

Point A does not move, but the rest of △II is turned to align with △I. One solution is to rotate △II **90° counterclockwise.** The fact that the triangles perfectly match after the rotation shows that they are congruent.

Conclude Describe an alternate way to rotate △II to the position of △I.

Activity 2

Rotations and Congruence

Materials needed:
- scissors
- unlined paper

One way to show that two figures are congruent is to move one figure to the position of the other figure to see if the two figures perfectly match.

a. **Model** Fold a sheet of paper in half and cut a shape from the doubled sheet of paper so that two congruent shapes are cut out at the same time. Then position the two figures on your desk so that a rotation is the only movement necessary to move one shape onto the other shape. Perform the rotation to show that the shapes are congruent.

b. **Represent** On another sheet of paper, draw or trace the two shapes you cut out. Draw the shapes in such a position that a 90° rotation of one shape would move it to the position of the other shape.

Lesson Practice

a. **Predict** Wakeisha skated east, turned 180° clockwise, and then continued skating. In what direction was Wakeisha skating after the turn?

Describe each rotation in degrees clockwise or counterclockwise:

b. a quarter turn to the left

c. a full turn to the right

d. a three quarter turn to the left

e. a half turn to the right

f. Describe the rotation about point *A* that would position triangle 1 on triangle 2.

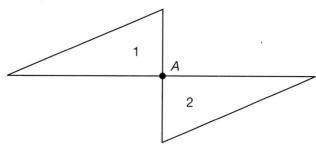

*1. If it is not a leap year, what is the total number of days in January,
(RF12) February, and March?

2. A tailor made each of 12 children a pair of pants and 2 shirts. How
(39) many pieces of clothing did the tailor make?

3. Burke did seven more chin-ups than Ariel did. If Burke did eighteen
(16) chin-ups, how many chin-ups did Ariel do?

4. Kadeeja drove 200 miles on 8 gallons of gas. Her car averaged how
(60) many miles on each gallon of gas?

*5. Melinda paid the clerk $20.00 for a book that was priced at $8.95. The
(28, 52) tax was 54¢. How much money should she get back?

*6. a. Which two prime numbers are factors of 15?
(56)

   b. **Explain** Is 15 a prime number? Why or why not?

7. If each side of an octagon is 1 centimeter long, what is the octagon's
(20, 42, 50) perimeter in millimeters?

8. **Represent** One third of the 18 marbles were blue. How many
(74) of the marbles were blue? Draw a picture to solve the problem.

*9. a. **Analyze** The Mendez family hiked 15 miles in 1 day. At that rate,
(58, Inv. 7) how many miles would they hike in 5 days? Make a table to solve
   the problem.

   b. **Formulate** Write an equation to represent the data in the table.

*10. **Explain** Mylah picked 364 peaches in 7 days. She picked an
(53, 69) average of how many peaches each day? Explain why your answer is
   reasonable.

*11. a. (**Analyze**) Zachary did 100 push-ups last week. He did 59 of those
(Inv. 4, 77) push-ups last Wednesday. What fraction of the 100 push-ups did
Zachary do last Wednesday?

b. (**Represent**) Write the answer to part **a** as a decimal number. Then
use words to name the number.

*12. (**Generalize**) Use this table to write a formula that can be used to
(73) convert any number of milliliters to liters. Use *l* for liters and *n* for
milliliters.

Liters	1	2	3	4	5
Milliliters	1000	2000	3000	4000	5000

13. $4.56 - (2.3 + 1.75)$
(9, 45)

14. $\sqrt{36} + n = 7 \times 8$
(Inv. 3, 64)

15. $3 \times 6 \times 3^2$
(65)

16. $462 \times \sqrt{9}$
(Inv. 3, 59)

17. $7^2 - \sqrt{49}$
(Inv. 3, 65)

18.
(71)
$$\begin{array}{r} 36 \\ \times\ 50 \\ \hline \end{array}$$

19.
(59)
$$\begin{array}{r} \$4.76 \\ \times\quad 7 \\ \hline \end{array}$$

20.
(4)
$$\begin{array}{r} 4 \\ 3 \\ 2 \\ 7 \\ 6 \\ 8 \\ +\ n \\ \hline 47 \end{array}$$

21. $\dfrac{114}{2}$
(69)

*22. $5\overline{)182}$
(72)

*23. $2\overline{)161}$
(75)

24. $2n = \$110$
(34, 75)

25. $5\overline{)400}$
(75)

*26. $\dfrac{490}{7}$
(75)

*27. Write 0.32 as a fraction using words.
(Inv. 4)

28. Find the perimeter and area of this square:
(66)

3 yards

3 yards ☐

*29. **(Represent)** Draw the capital letter *E* rotated 90° clockwise.
(78)

E

30. **(Estimate)** Which angles in this figure look like right angles?
(17)

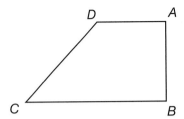

Real-World Connection

Christina earns money by doing different kinds of work for her neighbor, Mrs. Long. Refer to the tables below.

Money Earned by Babysitting	
Hours	**Dollars**
1	5.50
2	11
3	16.50

Money Earned by Dog Walking	
Dogs	**Dollars**
1	1.25
2	2.50
3	3.75

Money Earned by Car Washing	
Cars	**Dollars**
1	7
2	14
3	21

a. If Christina baby sits for Mrs. Long for 4 hours, how much money will she earn?

b. Mrs. Long owns 2 cars. If Christina washes Mrs. Long's cars twice this month, how much will she earn?

c. Mrs. Long owns 1 dog. If Christina walks Mrs. Long's dog every day, how much will she earn in 1 week?

California Mathematics Content Standards

NS **3.0, 3.2** Demonstrate an understanding of, and the ability to use, standard algorithms for multiplying a multidigit number by a two-digit number and for dividing a multidigit number by a one-digit number; use relationships between them to simplify computations and to check results.

NS **3.0, 3.4** Solve problems involving division of multidigit numbers by one-digit numbers.

• Division with Three-Digit Answers

Power Up

facts	Power Up G
count aloud	Count by fives from 4 to 54.
mental math	Before adding, make one number larger and the other number smaller in problems **a–c**.

    a. Number Sense: $48 + 37$

    b. Number Sense: $62 + 29$

    c. Number Sense: $135 + 47$

    d. Fractional Part: $\frac{1}{2}$ of \$20

    e. Fractional Part: $\frac{1}{4}$ of \$20

    f. Fractional Part: $\frac{1}{10}$ of \$20

    g. Estimation: Mario earns \$8.95 for each hour he works. About how much does Mario earn for working 6 hours?

    h. Calculation: $\sqrt{64}$, $\times\, 3$, $+\, 1$, $\times\, 2$, $+\, 98$

problem solving	Choose an appropriate problem-solving strategy to solve this problem. The bar graph below shows the number of students in each of the three fourth grade classes at Mayfair School. If seven new fourth graders were to start attending the school, how could they be assigned to the classes to make each class equal in size?

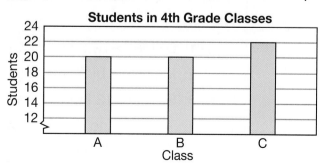

New Concept

We have practiced division problems that have two-digit answers. In this lesson we will practice division problems that have three-digit answers. Remember that the pencil-and-paper method we have used for dividing has four steps.

Step 1: Divide.   ÷

Step 2: Multiply.   ×

Step 3: Subtract.   −

Step 4: Bring down.   ↓

For each step we write a number. When we finish Step 4, we go back to Step 1 and repeat the steps until no digits remain to bring down.

Example 1

Divide: 3)794

Step 1: Divide 3)7 and write "2."

Step 2: Multiply 2 by 3 and write "6."

Step 3: Subtract 6 from 7 and write "1."

Step 4: Bring down the 9 to make 19.

Repeat:

Step 1: Divide 19 by 3 and write "6."

Step 2: Multiply 6 by 3 and write "18."

Step 3: Subtract 18 from 19 and write "1."

Step 4: Bring down the 4 to make 14.

Repeat:

Step 1: Divide 14 by 3 and write "4."

Step 2: Multiply 4 by 3 and write "12."

Step 3: Subtract 12 from 14 and write "2."

Step 4: There are no digits to bring down.
We are finished dividing. We write "2" as the remainder for a final answer of **264 R 2.**

Thinking Skills

(Discuss)

Why do we write the digit 2 in the hundreds place of the quotient?

$$
\begin{array}{r}
264 \text{ R } 2 \\
3\overline{)794} \\
\underline{6} \\
19 \\
\underline{18} \\
14 \\
\underline{12} \\
2
\end{array}
$$

Check:

$$
\begin{array}{r}
264 \\
\times 3 \\
\hline
792
\end{array}
$$

$$
\begin{array}{r}
792 \\
+ 2 \\
\hline
794
\end{array}
$$

To divide dollars and cents by a whole number, we divide the digits just like we divide whole numbers. *The decimal point in the answer is placed directly above the decimal point inside the division box.* We write a dollar sign in front of the answer.

Example 2

The total cost of three identical items is $8.40. What is the cost of each item?

The decimal point in the quotient is directly above the decimal point in the dividend. We write a dollar sign in front of the quotient.

The cost of each item is **$2.80.**

$$
\begin{array}{r}
\$2.80 \\
3\overline{)\$8.40} \\
\underline{6} \\
2\,4 \\
\underline{2\,4} \\
00 \\
\underline{00} \\
0
\end{array}
$$

Example 3

A local company is providing 4245 hats for a town festival. There are 5 different colors of hats and an equal number of each color. How many hats of each color are there?

We divide four-digit numbers the same way we divide three-digit numbers.

First we divide $5\overline{)42}$ Since we are dividing hundreds, we write 8 in the hundreds place of the quotient. Then, we subtract and bring down the 4. We continue to divide, multiply, subtract, and bring down.

There are **849 hats** of each color.

$$
\begin{array}{r}
849 \\
5\overline{)4245} \\
\underline{40} \\
24 \\
\underline{20} \\
45 \\
\underline{45} \\
0
\end{array}
$$

Lesson Practice

a. Copy the diagram at right. Then name the four steps of pencil-and-paper division.

Divide:

b. $4\overline{)974}$

c. $\$7.95 \div 5$

d. $6\overline{)1512}$

e. $8\overline{)\$50.00}$

*1. **Interpret** Use the information in the graph below to answer
(Inv. 5) parts **a–c.**

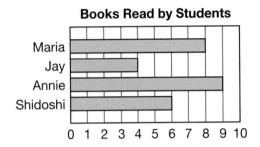

Books Read by Students

a. Which student has read exactly twice as many books as Jay?

b. Shidoshi's goal is to read 10 books. How many more books does
he need to read to reach his goal?

c. If the books Annie has read have an average of 160 pages each,
about how many pages has she read?

*2. Dala saw some pentagons. The pentagons had a total of 100 sides.
(50, 53) How many pentagons did Dala see?

*3. Mariah bought a rectangular piece of land that was 3 miles long and
(Inv. 3) 2 miles wide. She plans to divide the land into two sections of equal
area. If she farms one of the sections how many square miles could be
farmed?

*4. Max bought 10 pencils for 24¢ each. The tax was 14¢. What was the
(28, 37) total cost of the pencils?

5. **Multiple Choice** A full pitcher of orange juice contains about how
(73) much juice?

    A 2 ounces   **B** 2 liters   **C** 2 gallons   **D** 2 cups

6. **Represent** Draw a triangle that has two perpendicular sides. What
(17) type of angles did you draw?

7. (74) **Represent** One fourth of the 48 gems were rubies. How many of the gems were rubies? Draw a picture to solve the problem.

*8. a. (Inv. 4, 77) **Represent** One hundred fans attended the game, but only 81 fans cheered for the home team. What fraction of the fans who attended the game cheered for the home team?

b. Write the answer in part **a** as a decimal number. Then use words to name the number.

9. (45) $46.01 - (3.68 + 10.2)$

10. (8, 52) $728 + c = 1205$

11. (71) 36×40

12. (71) 20×42

13. (71) $\$2.75 \times 10$

14. (59)
$$\begin{array}{r} \$3.17 \\ \times \quad 4 \\ \hline \end{array}$$

15. (59)
$$\begin{array}{r} 206 \\ \times \quad 5 \\ \hline \end{array}$$

16. (71)
$$\begin{array}{r} 37 \\ \times \quad 40 \\ \hline \end{array}$$

17. (79) $3\overline{)492}$

18. (79) $5\overline{)860}$

19. (34, 79) $6m = \$9.30$

20. (65, 69) $168 \div 2^3$

*21. (75) $240 \div 4$

*22. (75) $241 \div 8$

23. (66) Find the perimeter and area of this rectangle:

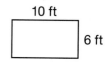
10 ft
6 ft

*24. (78) **Verify** Which of these letters will look the same if it is turned a half turn?

H A P P Y

*25. (17) **Estimate** Which angle in this figure looks like an obtuse angle?

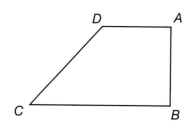

*26. **Multiple Choice** Matthew borrowed $12 from his sister, Lisa. He paid
  (76) her back $4. Lisa kept a record of the money Matthew owes her. Which
        number represents Matthew's balance with Lisa?

   A $16   **B** –$12   **C** –$8   **D** –$4

*27. The table shows the relationship between meters and centimeters:
  (42,
  Inv. 7)

Number of Meters	1	2	3	4	5
Number of Centimeters	100	200	300	400	500

   a. **Formulate** Write a formula to represent the relationship.

   b. **Predict** Use your formula to find the number of centimeters in
        10 meters.

28. In Dodge City, Kansas, the record high temperature in January, 1989
  (21) was 80°F. The record low temperature in January, 1984 was –13°F. How
        many degrees difference is this?

29. The Peace River is 1210 miles long, and its source is in British Columbia.
  (27) The Red River is 1290 miles long, and its source is in New Mexico.
        Which river is longer?

*30. **Model** Draw a number line from 6 to 7 divided into tenths. On it show
  (43) the locations of 6.1, $6\frac{3}{10}$, 6.6, and $6\frac{9}{10}$.

LESSON 80

California Mathematics Content Standards

NS 2.0, 2.2 Round two-place decimals to one decimal or the nearest whole number and judge the reasonableness of the rounded answer.

AF 1.0, 1.1 Use letters, boxes, or other symbols to stand for any number in simple expressions or equations (e.g., demonstrate an understanding and the use of the concept of a variable).

AF 1.0, **1.5** Understand that an equation such as $y = 3x + 5$ is a prescription for determining a second number when a first number is given.

• Mass and Weight

facts   Power Up G

count aloud   Count by fourths from $2\frac{1}{2}$ to $7\frac{1}{2}$.

mental math   Subtracting two-digit numbers mentally is easier if the second number ends in zero. By increasing both numbers in a subtraction by the same amount, we can sometimes make the subtraction easier while keeping the difference the same. For example, instead of $45 - 28$, we can think $47 - 30$. We added two to 28 to make it end in zero and then added two to 45 to keep the difference the same. Use this strategy in problems **a–d.**

    a. Number Sense: $45 - 39$

    b. Number Sense: $56 - 27$

    c. Number Sense: $63 - 48$

    d. Number Sense: $82 - 35$

    e. Powers/Roots: Compare: $\sqrt{16} - \sqrt{9} \bigcirc 1^2$

    f. Measurement: The high temperature was 84°F. The low temperature was 68°F. The difference between the high and low temperatures was how many degrees?

    g. Estimation: Each candle costs \$3.05. If Miranda has \$12, does she have enough to buy 4 candles?

    h. Calculation: $\frac{1}{4}$ of 24, × 9, − 15, + 51

problem solving   Choose an appropriate problem-solving strategy to solve this problem. Tahlia's soccer team, the Falcons, won their match against the Eagles. There were 11 goals scored altogether by both teams. The Falcons scored 3 more goals than the Eagles. How many goals did each team score?

There is a difference between *weight* and *mass*. The **mass** of an object is how much matter an object has within itself. **Weight** is the measure of the force of gravity on that object. Though an object's weight depends on the force of gravity, its mass does not. For example, the force of gravity on the moon is less than it is on Earth, so the weight of an object on the moon is less, but its mass remains the same.

The units of weight in the U.S. Customary System are **ounces, pounds,** and **tons.** We can use the word ounce to describe an amount of fluid. However, ounce can also describe an amount of weight. A fluid ounce of water weighs about one ounce.

As we see in the table below, one *pound* is 16 ounces, and one *ton* is 2000 pounds. Ounce is abbreviated oz. Pound is abbreviated lb.

$$16 \text{ oz} = 1 \text{ lb}$$
$$2000 \text{ lb} = 1 \text{ ton}$$

A box of cereal might weigh 24 ounces. Some students weigh 98 pounds. Many cars weigh 1 ton or more.

24 ounces   98 pounds   1 ton

Example 1

This book weighs about 2 pounds. Two pounds is how many ounces?

Each pound is 16 ounces. So 2 pounds is 2 × 16 ounces, which is **32 ounces.**

Example 2

The rhinoceros weighed 3 tons. Three tons is how many pounds?

Each ton is 2000 pounds. This means 3 tons is 3 × 2000 pounds, which is **6000 pounds.**

Grams and *kilograms* are metric units of mass. Recall that the prefix *kilo–* means "thousand." So a kilogram is 1000 grams. Gram is abbreviated g. Kilogram is abbreviated kg.

$$1000 \text{ g} = 1 \text{ kg}$$

A dollar bill has a mass of about 1 gram. This book has a mass of about 1 kilogram. Since this book has fewer than 1000 pages, each page is more than 1 gram.

Example 3

Choose the more reasonable measure for a–c.

   a. pair of shoes: 1 g or 1 kg

   b. cat: 4 g or 4 kg

   c. quarter: 5 g or 5 kg

   a. A pair of shoes is about **1 kg.**

   b. A cat is about **4 kg.**

   c. A quarter is about **5 g.**

Example 4

How many grams are equal to 5 kilograms?

We can find the answer by setting up a table.

Kilograms	1	2	3	4	5
Grams	1000	2000	3000	4000	5000

5 kilograms = **5000 grams**

Write can write a formula for converting kilograms to grams. We use *g* for grams and *k* for kilograms and write $g = 1000k$ or $1000k = g$. We can use this formula to convert any number of kilograms to grams.

Generalize How could we rewrite the formula $g = 1000k$ using *x* and *y*?

Example 5

> **Mr. and Mrs. Gordon purchased two salads at the salad bar. The salads weighed 1.43 lb and 1.37 lb. To the nearest tenth of a pound, how much did the salads weigh?**
>
> We are asked to round each number to the nearest tenth.
>
> 1.43 lb rounds to 1.4 lb
>
> 1.37 lb rounds to 1.4 lb
>
> Now we can add. 1.4 lb + 1.4 lb = **2.8 lb**
>
> (**Discuss**) The restaurant charges $2 per pound for a salad. The cashier will round up to the nearest pound. How much did the Gordon's pay for their salads?

Lesson Practice

  a. Dave's pickup truck can haul a half ton of cargo. How many pounds is a half ton?

  b. The newborn baby weighed 7 lb 12 oz. The baby's weight was how much less than 8 pounds?

(**Estimate**) Choose the more reasonable measure in problems **c–e.**

  c. tennis ball: 57 g or 57 kg

  d. dog: 6 g or 6 kg

  e. bowling ball: 7 g or 7 kg

  f. Seven kilograms is how many grams?

  g. Which depends on the force of gravity: mass or weight?

Written Practice — *Distributed and Integrated*

1. It takes Tempest 20 minutes to walk to school. At what time should she
(13) start for school if she wants to arrive at 8:10 a.m.?

*2. A container and its contents weigh 125 pounds. The contents of the
(10, 12) container weigh 118 pounds. What is the weight of the container?

*3. Anjelita is shopping for art supplies and plans to purchase a sketchpad
(12, 28) for $4.29, a charcoal pencil for $1.59, and an eraser for 69¢. If the amount of sales tax is 43¢ and Anjelita pays for her purchase with a $10 bill, how much change should she receive?

4. According to this calendar, October 30, 1904,
(RF12) was what day of the week?

| OCTOBER 1904 |
S	M	T	W	T	F	S
						1
2	3	4	5	6	7	8
9	10	11	12	13	14	15
16	17	18	19	20	21	22
23	24	25	26	27	28	29
30	31					

*5. **Explain** From 3:00 p.m. to 3:45 p.m., the minute hand of a clock
(78) turns how many degrees? Explain your thinking.

6. Round three thousand, seven hundred eighty-two to the nearest
(27, 40) thousand.

7. The limousine weighed 2 tons. How many pounds is 2 tons?
(80)

8. **Represent** One fifth of the 45 horses were pintos. How
(74) many of the horses were pintos? Draw a picture to illustrate
the problem.

*9. **Analyze** What fraction of the set of triangles is shaded?
(77)

10. **Represent** Which point on the number line below could represent
(Inv. 2, 27) 23,650?

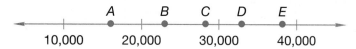

*11. **Connect** Write each decimal number as a fraction:
(Inv. 4)
    a. 0.1   **b.** 0.01   **c.** 1.11

12. $36.47
(28, 51) + $ 9.68

13. $30.00
(28, 52) − $13.45

14.   6
(11)   8
    17
    23
  110
    25
+ 104

15.   476
(59) ×   7

16.   804
(59) ×   5

17. $12.65 - (7.43 - 2.1)$
(9, 45)

18. $5^2 + 5^2 + n = 10^2$
(8, 65)

19. (Represent) Write each of these numbers with words:
(Inv. 4)
     a. $2\frac{1}{10}$ **b.** 2.1

*** 20.** $10 \times 23¢$ *** 21.** 62×30 *** 22.** $70 \times \$25$
(71)   (71)   (71)

*** 23.** $3)\overline{\$6.27}$ **24.** $7)\overline{820}$ **25.** $6)\overline{333}$
(79)   (79)   (72)

26. $625 \div \sqrt{25}$ *** 27.** $400 \div 2^3$ **28.** $2w = 1370$
(Inv. 3,   (65, 75)   (34, 79)
79)

29. Find the perimeter and area of this square.
(66)

                                                     10 m

*** 30.** (Represent) The table below is based on a survey about favorite
(Inv. 6, flowers. Use the data in the table to make a double bar graph. Write
76) one conclusion based on the graph.

Flower	Adults	Children
Rose	6	2
Daisy	4	9
Mum	5	4

California Mathematics Content Standards

NS 1.0, 1.6 Write tenths and hundredths in decimal and fraction notations and know the fraction and decimal equivalents for halves and fourths (e.g., $\frac{1}{2} = 0.5$ or .50; $\frac{7}{4} = 1\frac{3}{4} = 1.75$).

NS 1.0, 1.7 Write the fraction represented by a drawing of parts of a figure; represent a given fraction by using drawings; and relate a fraction to a simple decimal on a number line.

MR 2.0, 2.3 Use a variety of methods, such as words, numbers, symbols, charts, graphs, tables, diagrams, and models, to explain mathematical reasoning.

Focus on

Investigating Equivalent Fractions with Manipulatives

Fraction manipulatives can help us better understand fractions. In this investigation we will make and use a set of fraction manipulatives.

Activity 1

Using Fraction Manipulatives

Materials needed:
- **Lesson Activities 11, 12,** and **13**
- scissors
- envelopes or locking plastic bags (optional)

Model Use your fraction manipulatives to complete the following exercises:

1. Another name for $\frac{1}{4}$ is a quarter. How many quarters of a circle does it take to form a whole circle? Show your work.

2. Fit two quarter circles together to form a half circle. That is, show that $\frac{2}{4}$ equals $\frac{1}{2}$.

3. How many fourths equals $1\frac{1}{4}$?

4. This number sentence shows how to make a whole circle using half circles:

$$\frac{1}{2} + \frac{1}{2} = 1$$

Write a number sentence that shows how to make a whole circle using only quarter circles.

5. How many half circles equals $1\frac{1}{2}$ circles?

6. Four half circles make how many whole circles?

Model Manipulatives can help us compare and order fractions. Use your fraction manipulatives to illustrate and answer each problem:

7. Arrange $\frac{1}{2}$, $\frac{1}{8}$, and $\frac{1}{4}$ in order from least to greatest.

8. Arrange $\frac{3}{8}$, $\frac{3}{4}$, and $\frac{1}{2}$ in order from greatest to least.

9. $\frac{2}{2}$ ◯ $\frac{2}{4}$
10. $\frac{4}{8}$ ◯ $\frac{3}{8}$

11. **Generalize** If the denominators of two fractions are the same, how can we determine which fraction is larger and which is smaller?

12. **Generalize** If the numerators of two fractions are the same, how can we determine which fraction is larger and which is smaller?

Manipulatives can also help us **reduce** fractions. When we reduce a fraction, we do not change the size of the fraction. We just use smaller numbers to name the fraction. (With manipulatives, we use fewer pieces to form the fraction.) For example, we may reduce $\frac{2}{4}$ to $\frac{1}{2}$. Both $\frac{2}{4}$ and $\frac{1}{2}$ name the same portion of a whole, but $\frac{1}{2}$ uses smaller numbers (fewer pieces) to name the fraction.

Use your fraction manipulatives to help you reduce the fractions in problems **13–16**. Show how the two fractions match.

13. $\frac{2}{4}$
14. $\frac{2}{8}$

15. $\frac{4}{8}$
16. $\frac{6}{8}$

Manipulatives can also help us add and subtract fractions. Illustrate each addition below by combining fraction manipulatives. Record each sum.

17. $\frac{1}{4} + \frac{2}{4}$
18. $\frac{2}{8} + \frac{3}{8}$

To illustrate each subtraction in problems **19–21**, form the first fraction; then separate the second fraction from the first fraction. Record what is left of the first fraction as your answer.

19. $\frac{3}{4} - \frac{2}{4}$
20. $\frac{4}{8} - \frac{1}{8}$

21. $\frac{2}{2} - \frac{1}{2}$

Activity 2

Understanding How Fractions and Decimals are Related

Fraction manipulatives can help us understand how fractions and decimals are related. Use the decimal labels on your manipulatives to answer these problems:

22. One half of a circle is what decimal portion of a circle?

23. What decimal portion of a circle is $\frac{1}{4}$ of a circle?

24. What decimal portion of a circle is $\frac{3}{4}$ of a circle?

25. What decimal number is equivalent to $\frac{1}{3}$?

26. What decimal number is equivalent to $\frac{1}{5}$?

27. What decimal number is equivalent to $\frac{1}{8}$?

28. Compare: 0.125 $\bigcirc$ 0.25

29. Form a half circle using two $\frac{1}{4}$ pieces. Here is a fraction number sentence for the model:

$$\frac{1}{4} + \frac{1}{4} = \frac{1}{2}$$

Write an equivalent number sentence using the decimal numbers on the pieces.

30. Form $\frac{3}{4}$ of a circle two ways. First use three $\frac{1}{4}$ pieces. Then use a $\frac{1}{2}$ piece and a $\frac{1}{4}$ piece. Here are the two fraction number sentences for these models:

$$\frac{1}{4} + \frac{1}{4} + \frac{1}{4} = \frac{3}{4} \qquad \frac{1}{2} + \frac{1}{4} = \frac{3}{4}$$

Write equivalent number sentences using the decimal numbers on these pieces.

31. Form a whole circle using four $\frac{1}{4}$ pieces. Then take away one of the $\frac{1}{4}$ pieces. Here is a fraction number sentence for this subtraction. Write an equivalent number sentence using the decimal numbers on the pieces.

$$1 - \frac{1}{4} = \frac{3}{4}$$

32. Form a half circle using four $\frac{1}{8}$ pieces. Then take away one of the pieces. Here is a fraction number sentence for this subtraction. Write an equivalent number sentence using the decimal numbers on the pieces.

$$\frac{1}{2} - \frac{1}{8} = \frac{3}{8}$$

33. Here we show $\frac{3}{4}$ of a circle and $\frac{1}{2}$ of a circle:

We see that $\frac{3}{4}$ is greater than $\frac{1}{2}$. In fact, we see that $\frac{3}{4}$ is greater than $\frac{1}{2}$ by a $\frac{1}{4}$ piece. Here we show a larger-smaller-difference number sentence for this comparison:

$$\frac{3}{4} - \frac{1}{2} = \frac{1}{4}$$

Investigate Further

a. Parker wanted to make pumpkin bread following the recipe below:

Pumpkin Bread
1 (15 ounce) can pumpkin puree
1 cup vegetable oil
4 eggs
$3\frac{1}{2}$ cups all-purpose flour
3 cups white sugar
$1\frac{2}{4}$ teaspoons baking soda
$1\frac{1}{2}$ teaspoons salt
$1\frac{1}{2}$ teaspoons ground allspice
$1\frac{2}{4}$ teaspoons ground nutmeg
2 teaspoons ground cinnamon
$\frac{1}{2}$ cup chopped walnuts
1 teaspoon baking powder

Use your fraction manipulatives to find each amount.

How many cups will Parker place in a mixing bowl if he combines the oil, flour, sugar, and walnuts first?

If Parker combines the baking soda, salt, allspice, nutmeg, and cinnamon in another bowl, how much mixture will be in the bowl?

California Mathematics Content Standards

MG 3.0, 3.5 Know the definitions of a right angle, an acute angle, and an obtuse angle. Understand that 90°, 180°, 270°, and 360° are associated, respectively, with $\frac{1}{4}$, $\frac{1}{2}$, $\frac{3}{4}$, and full turns.

MG 3.0, 3.7 Know the definitions of different triangles (e.g., equilateral, isosceles, scalene) and identify their attributes.

MR 2.0, 2.4 Express the solution clearly and logically by using the appropriate mathematical notation and terms and clear language; support solutions with evidence in both verbal and symbolic work.

• Classifying Triangles

facts	Power Up I
count aloud	Count by fours from 80 to 120.
mental math	Find each difference by first increasing both numbers so that the second number ends in zero in problems **a–c.**

a. **Number Sense:** 63 – 28

b. **Number Sense:** 45 – 17

c. **Number Sense:** 80 – 46

d. **Money:** Noah had $10.00. Then he spent $5.85 on lunch. How much money did he have left over?

e. **Measurement:** How many inches is $\frac{1}{2}$ of a foot?

f. **Measurement:** How many inches is $\frac{1}{4}$ of a foot?

g. **Estimation:** The total cost for 4 movie rentals was $15.92. Round this amount to the nearest dollar and then divide by 4 to estimate the cost per rental.

h. **Calculation:** 5^2, × 2, × 2, × 2, × 2

problem solving

In the diagram at right, the circle stands for students who have one or more pets at home. A letter inside the circle stands for a particular student who has a pet. A letter outside the circle stands for a student who does not have any pets. The letter *A* stands for Adrian, who has a dog. The letter *B* stands for Brooke, who does not have any pets. Copy the graph on your paper. On the graph, place the letter *C* for Clarrisa, who keeps a goldfish, and *D* for David, who does not have pets.

Students with Pets

A

B

One way to classify (describe) a triangle is by referring to its largest angle as either obtuse, right, or acute. An obtuse angle is larger than a right angle. An acute angle is smaller than a right angle.

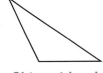
Obtuse triangle
(One angle is obtuse.)

Right triangle
(One angle is right.)

Acute triangle
(All angles are acute.)

Another way to classify a triangle is by comparing the lengths of its sides. If all three sides are equal in length, the triangle is **equilateral.** If at least two sides are equal in length, the triangle is an **isosceles triangle.** If all three sides have different lengths, the triangle is an **scalene triangle.**

Equilateral triangle

Isosceles triangle

Scalene triangle

Represent Can an isosceles triangle have an obtuse angle? Draw a triangle to support your conclusion.

Notice that the three angles of the equilateral triangle are the same size. This means an equilateral triangle is also **equiangular.** Now notice that only two angles of the isosceles triangle are the same size. In a triangle, the number of angles with the same measure equals the number of sides with the same measure.

Example 1

Draw a triangle that is both a right triangle and an isosceles triangle.

A right triangle contains one right angle. An isosceles triangle has two sides of equal length. We begin by drawing a right angle with equal-length sides.

Then we draw the third side of the triangle.

Classify Describe the angles of a right triangle.

Example 2

Describe this triangle in as many ways as possible.

Write can see that this triangle has all sides the same length so it is an **equilateral triangle.** Each angle is acute so it is an **acute triangle.** All three angles are the same number of degrees so it is an **equiangular triangle.** It has two or more sides of the same length, so it is also an **isosceles triangle.**

Conclude Can an isosceles triangle also be a scalene triangle? Why or why not?

Lesson Practice

a. **Conclude** Can a right triangle have two right angles? Why or why not?

b. What is the name for a triangle that has at least two sides equal in length?

c. Draw an acute scalene triangle.

Written Practice   *Distributed and Integrated*

* **1.** **Analyze** Use the following information to answer parts **a–c.**
(13, 76)

Freeman rode his bike 2 miles from his house to Didi's house. Together they rode 4 miles to the lake. Didi caught 8 fish. At 3:30 p.m. they rode back to Didi's house. Then Freeman rode home.

a. Altogether, how far did Freeman ride his bike?

b. It took Freeman an hour and a half to get home from the lake. At what time did he get home?

c. Didi caught twice as many fish as Freeman. How many fish did Freeman catch?

*2. **(Conclude)** Describe the number of degrees and the direction of a turn that would move this letter B to an upright position.
(78)

3. **(Estimate)** Find a reasonable sum of 4876 and 3149 by rounding each number to the nearest thousand and then adding.
(40, 61)

4. **(Explain)** What is the perimeter of a pentagon if each side is 20 centimeters long? Explain your reasoning.
(20, 50)

*5. **(Estimate)** Find the length of this segment to the nearest quarter inch:
(42)

```
inch        1           2           3           4
```

6. **(Represent)** One half of the 18 players were on the field. How many players were on the field? Draw a picture to illustrate the problem.
(74)

7. A dime is $\frac{1}{10}$ of a dollar. What fraction of a dollar is a penny?
(29)

*8. **(Generalize)** Write an equation to show how to convert any number of ounces to pounds. Use o for ounces and p for pounds.
(80)

*9. **(Analyze)** What fraction of the set of triangles is shaded?
(77)

△ △ △ △ △

10. **(Represent)** One millimeter is $\frac{1}{10}$ of a centimeter. Write that number as a decimal number. Then use words to write the number.
(Inv. 4)

*11.   31
(71)   × 20

12.   51
(71)   × 30

*13.   25
(71)   × 50

14. 7)1000
(79)

15. 3)477
(79)

16. 5)2585
(79)

17. $15.48 ÷ 9
(79)

18. 716 ÷ 4
(79)

19. 8x = 352
(34, 69)

***20.** **Multiple Choice** Which of the following is equivalent to $\frac{1}{2}$?
(Inv. 8)

A
$\frac{1}{3}$

B
$\frac{2}{6}$

C
$\frac{1}{4}$

D
$\frac{2}{4}$

21. (**Represent**) Use digits to write the number three million, eight hundred fifty
(47, 48) thousand in standard form. Then round it to the nearest million.

***22.** **a.** (**Evaluate**) If $y = 2x + 3$, then what is y when $x = 1$, 2, and 3?
(64,
Inv. 7)

b. (**Model**) Use **Lesson Activity 26** and graph $y = 2x + 3$ on a
coordinate graph.

23. Write the name of the property illustrated by each equation.
(RF9, 23,
26) **a.** $a + b = b + a$

b. $a(b \cdot c) = (a \cdot b) \cdot c$

c. $a \cdot 0 = 0$

d. $a + 0 = a$

***24.** Find the perimeter and area of this rectangle:
(66)

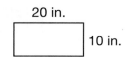

20 in.

10 in.

25. (**Represent**) Draw an equilateral triangle with sides 2 cm long.
(81)

26. What is the perimeter in millimeters of the triangle you drew in
(20, 42) problem **25?**

*27. a. (**Conclude**) In this polygon, which side appears to be parallel to side *AB*?

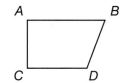

b. Which angle appears to be obtuse?

28. This graph shows the relationship between Colby's age and Neelam's age. How old was Neelam when Colby was 4 years old?

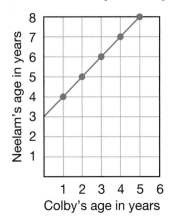

29. (**Represent**) Each grid represents a decimal number. Write each decimal number. Then write the sum and the difference of those numbers.

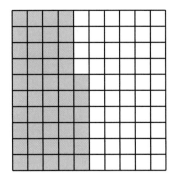
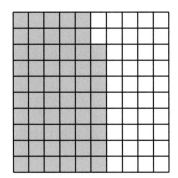

30. (**Estimate**) A mail carrier worked from 8 a.m. to noon and from 1 p.m. to 4 p.m. During those times, the carrier delivered mail to 691 homes. About how many deliveries did the carrier make each hour? Explain your answer.

California Mathematics Content Standards

MG 3.0, 3.4 Identify figures that have bilateral and rotational symmetry.

MG 3.0, 3.5 Know the definitions of a right angle, an acute angle, and an obtuse angle. Understand that 90°, 180°, 270°, and 360° are associated, respectively, with $\frac{1}{4}$, $\frac{1}{2}$, $\frac{3}{4}$, and full turns.

MR 3.0, 3.3 Develop generalizations of the results obtained and apply them in other circumstances.

• Symmetry

facts	Power Up I
count aloud	Count by fives from 3 to 63.
mental math	Before adding, make one number larger and the other number smaller in problems **a–c**.

    a. Number Sense: 38 + 46

    b. Number Sense: 67 + 24

    c. Number Sense: 44 + 28

    d. Number Sense: $3 \times 50 \times 10$

    e. Number Sense: Counting by fives from 5, every number Julio says ends in 0 or 5. If he counts by fives from 9, then every number he says ends in which digit?

    f. Geometry: The radius of the truck tire was 15 inches. The diameter of the tire was how many inches?

    g. Estimation: The total cost for 6 boxes of snack bars was $17.70. Round this amount to the nearest dollar and then divide by 6 to estimate the cost per box.

    h. Calculation: $\frac{1}{4}$ of 40, $\times$ 2, $\div$ 10, $\times$ 8, + 59

problem solving

Landon is packing a lunch for the park. He will take one bottle of water, a sandwich, and a fruit. He will choose either a ham sandwich or a peanut butter and jelly sandwich. For the fruit, Landon will choose an apple, an orange, or a banana. Make a tree diagram to find the possible combinations of lunches that Landon can pack. Then list each possible combination.

Thinking Skills

Discuss

Name several real-world examples of line symmetry.

In nature, we often find balance in the appearance and structure of objects and living things. For example, we see a balance in the wing patterns of moths and butterflies. We call this kind of balance **reflective symmetry,** or just **symmetry.**

The dashes across this drawing of a moth indicate a **line of symmetry.** The portion of the figure on each side of the dashes is the *mirror image* of the other side. If we stood a mirror along the dashes, the reflection in the mirror would appear to complete the figure.

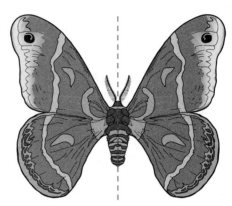

Some polygons and other figures have one or more lines of symmetry.

Example 1

Which of these polygons does *not* have a line of symmetry?

A □   B △   C ▱   D □

The rectangle has two lines of symmetry.

The isosceles triangle has one line of symmetry.

The square has four lines of symmetry.

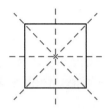

The third polygon has no line of symmetry. The answer is **C.**

(**Conclude**) Will every regular polygon always have at least one line of symmetry? Explain why or why not.

About half of the uppercase letters in the alphabet have lines of symmetry.

Example 2

Copy these letters and draw each line of symmetry, if any.

C H A I R

The letters **H** and **I** each have two lines of symmetry. The letters **C** and **A** each have one line of symmetry. The letter **R** has no lines of symmetry.

C H A I R

(**Represent**) Print the letters of your first name and describe any lines of symmetry those letters have.

Activity

Reflections and Lines of Symmetry

Materials needed:
- **Lesson Activity 24**
- mirror

Use a mirror to find lines of symmetry in the figures on **Lesson Activity 24.**

The symmetry illustrated in Examples 1 and 2 is reflective symmetry. Another type of symmetry is **rotational symmetry.** A figure has rotational symmetry if it matches its original position as it is rotated.

For example, a square has rotational symmetry because it matches itself every quarter turn (90°).

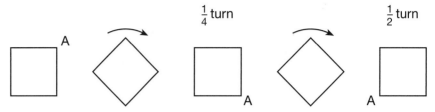

Likewise, the uppercase letter H has rotational symmetry because it matches its original position every half turn (180°).

Example 3

Which figures do *not* have rotational symmetry?

A

B

C

D

Figure **A** has rotational symmetry because it matches its original position every $\frac{1}{3}$ of a turn (120°).

Figure **B** has rotational symmetry because it matches its original position in one half of a turn (180°).

Figure **C** has rotational symmetry because it matches its original position every $\frac{1}{6}$ of a turn (60°).

Figure **D** does not have rotational symmetry because it requires a full turn (360°) to match its original position.

(**Discuss**) Draw a circle with one diameter. Does it have line symmetry? Does it have rotational symmetry?

Lesson Practice Copy each figure and draw the lines of symmetry, if any:

a.

b.

c.

d. W

e. X

f. Z

g. Which figures **a–f** do *not* have rotational symmetry?

h. Which figures in **a–f** have reflective symmetry?

i. Which of these polygons have reflective symmetry?

A

B

C

D

Written Practice *Distributed and Integrated*

1. How many kilograms are equal to 3000 grams?
(80)

*2. **Conclude** a. Which of these letters does *not* have a line
(82) of symmetry?

T N V W

b. Which of these letters has rotational symmetry?

Visit www.
SaxonMath.com/
Int4/ActivitiesCA
for an online
activity.

*3. Write the formula for the area of a square. Then find the area of
(66)   a square with sides 12 inches long.

4. Twenty-four inches is how many feet?
(42)

*5. a. Segment *YZ* is how many millimeters long?
(42)

  b. Segment *YZ* is how many centimeters long?

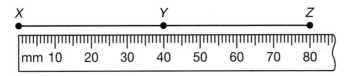

6. Jorge finished eating breakfast at the time shown on the
(13) clock. He finished eating lunch 5 hours 20 minutes later.
    What time did Jorge finish eating lunch?

7. (**Represent**) Write the number 7528 in expanded form. Then use
(27) words to write the number.

8. (**Represent**) One fifth of the 25 band members missed the note. How
(74) many band members missed the note? Draw a picture to illustrate the
    problem.

*9. Nikki cut a rectangular piece of paper along a diagonal to
(70, 78) make two triangles.

  a. How many degrees does Nikki need to turn one triangle
      so it will lay on top of the other triangle?

  b. Are the triangles congruent?

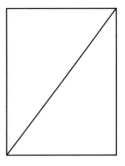

10. $6.35 + $14.25 + $0.97 + $5
(28)

11. 4.60 − (1.4 + 2.75)   **12.** $10.00 − (46¢ + $1.30)
(9, 45)   (9, 28)

*13. **Represent** Draw two rectangles that show that $\frac{1}{5} = \frac{2}{10}$.
(Inv. 8)

*14. **Analyze** Simplify each expression:
(63)
    a. $4 + 3 \times 3 - 1$ **b.** $16 \div 2 \times 4 + 5$

*15. **Verify** Is the product of two prime numbers prime? Why or
(56) why not?

*16. 28×20 ***17.**   13   **18.** $8.67
(71)   *(71)* $\underline{\times\ 30}$ *(59)* $\underline{\times\qquad 9}$

19. $7\overline{)3612}$ **20.** $6\overline{)\$33.30}$ **21.** $8\overline{)4971}$
(79)   *(79)*   *(79)*

22. $482 \div 5$ **23.** $270 \div 9$ **24.** $270 \div \sqrt{9}$
(72)   *(75)*   *(Inv. 3,*
                                                                  75)*

25. $7 + 7 + n = 7^2$ **26.** $3n = 6^2$
(64, 65)   *(34, 65)*

*27. a. **Represent** Draw an obtuse triangle.
(17, 81)

    b. 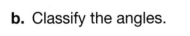 **Explain** Describe the segments of the obtuse angle. Explain
       your thinking.

28. The classroom was 42 feet long and 30 feet wide. How many
(Inv. 3, 1-foot square floor tiles were needed to cover the floor?
71)

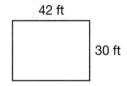

*29. a. **Classify** In polygon *ABCD*, which side appears to be
(17)   parallel to side *AD*?

    b. Classify the angles.

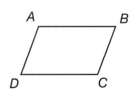

*30. **(Interpret)** This table shows the heights of several tall buildings.
(Inv. 5) Make a bar graph to display the data.

Tall Buildings in the United States

Building	Location	Height (stories)
The Pinnacle	Chicago, IL	48
Interstate Tower	Charlotte, NC	32
Two Union Square	Seattle, WA	56
28 State Street	Boston, MA	40

Early Finishers
Real-World Connection

Eduardo was given a sack of almonds. The label on the sack said it contained 60 almonds. He wanted to make sure, so he opened the sack and spread the almonds on a tabletop to count them. Eduardo recently learned about multiples at school, so he decided to count the almonds using multiples.

a. If Eduardo counts the nuts in groups of 10, what multiples will he name?

b. If he counts them in groups of 6, what multiples will he name?

c. If he counts the nuts in groups of 7, will Eduardo name 60 as a multiple?

California Mathematics Content Standards

NS 3.0, 3.4 Solve problems involving division of multidigit numbers by one-digit numbers.

AF 1.0, 1.1 Use letters, boxes, or other symbols to stand for any number in simple expressions or equations (e.g., demonstrate an understanding and the use of the concept of a variable).

• Division with Zeros in Three-Digit Answers

facts	Power Up I
count aloud	Count down by halves from 10 to $\frac{1}{2}$.
mental math	Before subtracting, make both numbers larger in problems **a–c.**

    a. Number Sense: $56 - 29$

    b. Number Sense: $43 - 18$

    c. Number Sense: $63 - 37$

    d. Money: Devin bought a vegetable tray for $7.52 and a bottle of fruit punch for $1.98. What was the total cost?

    e. Time: Compare: 72 hours $\bigcirc$ 2 days

    f. Time: About how many days are in 10 years?

    g. Estimation: The total cost for 3 boxes of cereal is $11.97. Round this amount to the nearest dollar and then divide by 3 to estimate the cost per box.

    h. Calculation: $\frac{1}{2}$ of 70, $\div$ 7, $\times$ 2, + 8, $\div$ 9, $\div$ 2

problem solving

Choose an appropriate problem-solving strategy to solve this problem. Twenty-one students went on the field trip to the zoo. Five students rode in the teacher's car. The rest of the students were divided equally among four cars driven by parents. How many students were in each of the cars driven by a parent? Explain how you found your answer.

Recall that the pencil-and-paper method we have used for dividing numbers has four steps:

Step 1: Divide.

Step 2: Multiply.

Step 3: Subtract.

Step 4: Bring down.

$\div$

$\times$

$-$

$\downarrow$

Every time we bring a number down, we return to Step 1. Sometimes the answer to Step 1 is zero, and we will have a zero in the answer.

Example 1

Each weekday afternoon in a small town, 618 newspapers are delivered to customers. The task of delivering the newspapers is divided equally among 3 drivers. How many newspapers does each driver deliver?

Thinking Skills

Verify

Why do we write the digit 2 in the hundreds place of the quotient?

Step 1: Divide $3\overline{)6}$ and write "2."

Step 2: Multiply 2 by 3 and write "6."

Step 3: Subtract 6 from 6 and write "0."

Step 4: Bring down the 1 to make 01 (which is 1).

$$\begin{array}{r} 2 \\ 3\overline{)618} \\ \underline{6} \\ 01 \end{array}$$

Repeat:

Step 1: Divide 3 into 01 and write "0."

Step 2: Multiply 0 by 3 and write "0."

Step 3: Subtract 0 from 1 and write "1."

Step 4: Bring down the 8 to make 18.

Thinking Skills

Discuss

Why do we write the digit 0 in the tens place of the quotient?

$$\begin{array}{r} 206 \\ 3\overline{)618} \\ \underline{6} \\ 01 \\ \underline{0} \\ 18 \\ \underline{18} \\ 0 \end{array}$$

Repeat:

Step 1: Divide 3 into 18 and write "6."

Step 2: Multiply 6 by 3 and write "18."

Step 3: Subtract 18 from 18 and write "0."

Step 4: There are no more digits to bring down, so the division is complete. The remainder is zero.

Each driver delivers **206 papers.**

Example 2

Divide: $4\overline{)1483}$

Step 1: Divide $4\overline{)14}$ and write "3."

Step 2: Multiply 3 by 4 and write "12."

Step 3: Subtract 12 from 14 and write "2."

Step 4: Bring down the 8 to make 28.

Repeat:

Step 1: Divide 4 into 28 and write "7."

Step 2: Multiply 7 by 4 and write "28."

Step 3: Subtract 28 from 28 and write "0."

Step 4: Bring down the 3 to make 03 (which is 3).

Repeat:

Step 1: Divide 4 into 03 and write "0."

Step 2: Multiply 0 by 4 and write "0."

Step 3: Subtract 0 from 3 and write "3."

Step 4: There are no digits to bring down, so the division is complete. We write "3" as the remainder.

$$
\begin{array}{r}
370\ \text{R}\ 3 \\
4\overline{)1483} \\
\underline{12} \\
28 \\
\underline{28} \\
03 \\
\underline{0} \\
3
\end{array}
$$

Justify How can we check the answer?

Example 3

The same number of landscaping bricks is stacked on each of 4 pallets. The total weight of the pallets is 3 tons. What is the weight in pounds of each pallet?

First we find the number of pounds in 3 tons. Each ton is 2 thousand pounds, so 3 tons is 6 thousand pounds. Now we find the weight of each pallet of bricks by dividing 6000 by 4.

We find that each pallet of bricks weighs **1500 pounds.**

$$
\begin{array}{r}
1500 \\
4\overline{)6000} \\
\underline{4} \\
20 \\
\underline{20} \\
000
\end{array}
$$

Lesson Practice

a. List the four steps of division and draw the division diagram.

Divide:

b. $4\overline{)815}$

c. $5\overline{)4152}$

Divide. Show your answer with a remainder.

d. $6\overline{)5432}$

e. $7\overline{)845}$

Divide mentally:

f. $5\overline{)1500}$ **g.** $4\overline{)2000}$

h. Find the missing factor in the equation $3m = 1200$.

Written Practice   *Distributed and Integrated*

*1. Mr. Carson bought 2 packages of turkey. They weighed 3.24 kg and
(62, 80)   2.26 kg. To the nearest tenth, how much did the packages weigh
altogether?

2. **Justify** On the package there were two 39¢ stamps, two 20¢
(28)   stamps, and one 15¢ stamp. Altogether, how much did the stamps
on the package cost? Explain why your answer is reasonable.

3. Daniella read 20 pages each day. How many pages did she read in
(39, 71)   2 weeks?

4. In the first track meet of the season, Wyatt's best triple jump measured
(42)   36 feet. What was the distance of that jump in yards?

5. What is the perimeter of this isosceles triangle in
(20, 42)   centimeters?

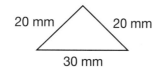

20 mm   20 mm

30 mm

*6. **Multiple Choice** Which of these tallies represents a prime number?
(56,
Inv. 6)   **A** ｜｜｜｜ ｜｜｜   **B** ｜｜｜｜ ｜｜｜｜

C ｜｜｜｜ ｜｜｜｜ ｜   **D** ｜｜｜｜ ｜｜｜｜ ｜｜

*7. **Multiple Choice** About how much liquid is in this medicine
(73)   dropper?

A 2 milliliters   **B** 2 liters

C 2 pints   **D** 2 cups

8. Solve for *n*: $87 + 0 = 87 \times n$
(64)

*9. **Represent** One third of the 24 students finished early. How
(74)   many students finished early? Draw a picture to illustrate the
problem.

*10. a. **Multiple Choice** Sketch each of the triangles below. Which of
(81)   these triangles does *not* exist?

   A a scalene right triangle   **B** an isosceles right triangle

   C an equilateral right triangle   **D** an equilateral acute triangle

   b. 🖉 **Justify** Explain why the triangle you chose does not
   exist.

11. $478.63
(28, 51)   + $ 32.47

12.   137,140
(52)   − 129,536

13. $60.00
(28, 52)   − $24.38

*14. 72 × 90
(71)

15.   28
(71)   × 50

*16.   25
(71)   × 40

17. $4.76
(59)   ×   8

*18.   210
(59)   ×   3

*19.   204
(59)   ×   5

20. 4)3000
(83)

21. $5n = 635$
(34, 79)

22. 7)426
(75)

23. 8)3614
(79)

24. $\dfrac{2736}{6}$
(79)

25. How much is one fourth of $10.00?
(74, 79, 83)

*26. a. **Conclude** Which of these letters has exactly one line of symmetry?
(82)

Q R H T

   b. Which of these letters has rotational symmetry?

*27. a. **Represent** Draw a rectangle that is 5 cm long and 4 cm wide.
(18, 20, Inv. 3)

   b. What is the perimeter and area of the rectangle you drew?

*28. a. **Conclude** In this polygon, which side appears to be
(17, 82)   parallel to side *BC?*

   b. Copy this figure and draw its line of symmetry.

   c. Does this figure have rotational symmetry?

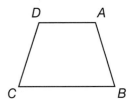

*29. **Analyze** Which two-digit number less than 20 is a multiple of both
(55)   4 and 6?

*30. **Interpret** This circle graph shows the results of an election for class
(Inv. 5)   president. Use the graph to answer the parts that follow.

Class Election Results

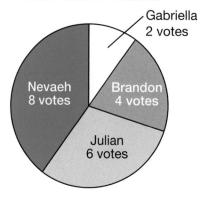

a. Which candidate won the election? How many votes did that
candidate receive?

b. Altogether, how many votes were cast in the election?

c. Which number is greater: the number of votes received by the
winner or the sum of the number of votes received by all of the
other candidates?

California Mathematics Content Standards

NS **3.0**, **3.3** Solve problems involving multiplication of multidigit numbers by two-digit numbers.

AF **1.0, 1.1** Use letters, boxes, or other symbols to stand for any number in simple expressions or equations (e.g., demonstrate an understanding and the use of the concept of a variable).

AF **1.0, 1.5** Understand that an equation such as $y = 3x + 5$ is a prescription for determining a second number when a first number is given.

• Multiplying by 10, 100, and 1000

facts	Power Up I
count aloud	Count down by quarters from 4 to $\frac{1}{4}$.
mental math	Counting by fives from 1, 2, 3, 4, or 5, we find five different final-digit patterns: 1 and 6; 2 and 7; 3 and 8; 4 and 9; and 5 and 0. When a number ending in 5 is added to or subtracted from another number, the final digit of that number and of the answer will fit one of the five patterns. Look for the final-digit patterns as you solve problems **a–f**.

  a. Number Sense: $22 + 5$

  b. Number Sense: $22 - 5$

  c. Number Sense: $38 + 5$

  d. Number Sense: $38 - 5$

  e. Number Sense: $44 + 5$

  f. Number Sense: $44 - 5$

  g. Estimation: Estimate the fraction of this circle that is shaded:

  h. Calculation: $\sqrt{36}$, $\times\ 3$, $+\ 10$, $\div\ 4$, $-\ 1$, $\div\ 3$

problem solving	Choose an appropriate problem-solving strategy to solve this problem. Tanner has three homework assignments to complete. One assignment is in math, one is in science, and one is in vocabulary. Tanner plans to finish one assignment before starting the next. What are the possible sequences in which he could complete the assignments?

New Concept

To multiply a whole number by 10, we simply add a zero to the end of the number.

$$
\begin{array}{r}
123 \\
\times\ \ \ 10 \\
\hline
1230
\end{array}
$$

When we multiply a whole number by 100, we add two zeros to the end of the number.

$$
\begin{array}{r}
123 \\
\times\ \ \ 100 \\
\hline
12{,}300
\end{array}
$$

When we multiply a whole number by 1000, we add three zeros to the end of the number.

$$
\begin{array}{r}
123 \\
\times\ \ \ 1000 \\
\hline
123{,}000
\end{array}
$$

When we multiply dollars and cents by a whole number, we remember to insert the decimal point two places from the right side of the product.

$$
\begin{array}{r}
\$1.23 \\
\times\ \ \ 100 \\
\hline
\$123.00
\end{array}
$$

Thinking Skills

Generalize

If we were to multiply 15 by 1 million, how many zeros would we attach to the right of the product of 15 and 1?

Example 1

Thinking Skills

Discuss

Why is the product of 100 and $6.12 *not* written as $6.1200?

Multiply mentally:

a. 37 × 10   b. $6.12 × 100   c. 45¢ × 1000

  a. The answer is "37" with one zero at the end:

  370

  b. The answer is "612" with two zeros at the end. We remember to place the decimal point and dollar sign:

  $612.00

  c. The answer is "45" with three zeros at the end. This makes 45,000¢, which in dollar form is

  $450.00

Example 2

A cement company delivered 10 tons of cement to a construction site. How many pounds is that?

We know that 2000 lbs = 1 ton. We can multiply 2000 × 10 to find the number of pounds.

$$2000 \times 10 = 20{,}000$$

The cement company delivered in **20,000 lbs** of cement.

Generalize Write a formula for changing any number of tons to pounds. Use p for pounds and t for tons. Find the number of pounds in 7 tons.

Lesson Practice Multiply mentally:

a. 365 × 10   **b.** 52 × 100   **c.** 7 × 1000

d. $3.60 × 10   **e.** 420 × 100   **f.** $2.50 × 1000

g. The table below shows the relationship between dimes and dollars. Write a formula to represent the relationship where d = dimes and l = dollars.

Number of Dollars	1	2	3	4	5
Number of Dimes	10	20	30	40	50

Written Practice *Distributed and Integrated*

1.
(15, 16, Inv. 5)
Interpret The line graph shows the average monthly temperatures during spring in Jacksonville, Florida. Use the graph to answer the parts that follow:

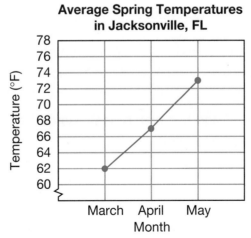

Average Spring Temperatures in Jacksonville, FL

a. What is the average temperature during March in Jacksonville, Florida? During April? During May?

b. Write a sentence that compares the average March temperature to the freezing temperature of water.

c. In Salt Lake City, Utah, the average May temperature is 14 degrees cooler than the average May temperature in Jacksonville, Florida. What is the average May temperature in Salt Lake City?

2. The 3-pound melon cost $1.44. What was the cost per pound?
(53, 72)

3. Jin spun all the way around in the air and dunked the basketball. Jin turned about how many degrees?
(78)

*4. Shunsuke bought a pair of shoes priced at $47.99. The sales tax was $2.88. Shunsuke gave the clerk $60.00. How much change should he receive?
(28, 51, 52)

5. (Analyze) If the perimeter of a square is 1 foot, how many inches long is each side?
(Inv. 3)

*6. The mass of a dollar bill is about 1 gram. Use this information to estimate the number of dollar bills it would take to equal 1 kilogram.
(80)

7. (Represent) One fourth of the 64 balloons were red. How many balloons were red? Draw a picture to illustrate the problem.
(74)

*8. a. T'Marra knew that her trip would take about 7 hours. If she left at half past nine in the morning, around what time should she arrive?
(13, 60, Inv. 7)

b. If T'Marra traveled 350 miles in 7 hours, then she traveled an average of how many miles each hour?

c. Using your answer to part **b,** make a table to show how far T'Marra would travel at her average rate in 1, 2, 3, and 4 hours.

*9. (Formulate) The product of 7 and 8 is how much greater than the sum of 7 and 8? Write an equation to show your work.
(63)

10. Compare: 3049 ◯ 3049.0
(44)

*11. **Estimate** Shakura purchased a birthday present for each of two
(46)   friends. Including sales tax, the cost of one present was \$16.61 and
       the cost of the other present was \$14.37. What is a reasonable estimate
       of the total cost of the presents? Explain your answer.

*12. What is the product of the fifth multiple of 2 and the eighth multiple of 6?
(55, 84)

*13. **Represent** Draw and shade two circles to show that $\frac{3}{4} = \frac{6}{8}$.
(Inv. 8)

*14. **Multiple Choice** Which of these words does *not* describe triangles
(81)   *ABC* and *DEF*?

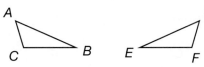

    A similar   **B** obtuse   **C** scalene   **D** isosceles

15. Find $0.625 - (0.5 + 0.12)$. Describe the steps in order.
(9, 45)

16. Mentally find this product. 47×100
(84)

17. $\begin{array}{r} 328 \\ \times\quad 4 \\ \hline \end{array}$
(59)

*18. $\begin{array}{r} 43 \\ \times\ 30 \\ \hline \end{array}$
(71)

*19. $\begin{array}{r} 35 \\ \times\ 40 \\ \hline \end{array}$
(71)

20. $5\overline{)4317}$
(79)

21. $8\overline{)\$40.00}$
(79, 83)

22. $6\overline{)3963}$
(83)

23. $3a = 426$
(34, 79)

24. $2524 \div 4$
(79)

*25. 60×100
(84)

*26. **Conclude** Below we show an equilateral triangle, an isosceles triangle, and a
(82)   scalene triangle. Name the triangle that does not have reflective symmetry.

27. $4 + 3 + 27 + 35 + 8 + n = 112$
(4)

* **28.** **a.** Segment *BC* is 1.7 cm long. How many centimeters long is
(42, 45)   segment *AB*?

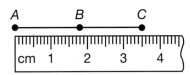

b. Write a decimal addition problem that is illustrated by the lengths of
segments *AB, BC,* and *AC.*

* **29.** **a.** Name a pair of parallel edges in the figure at right.
(17)

b. Name a pair of perpendicular edges.

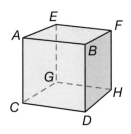

* **30.** **Explain** What is the length of segment *RT*? Explain your reasoning.
(Inv. 7)

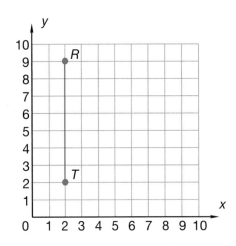

California Mathematics Content Standards

NS 1.0, **1.2** Order and compare whole numbers and decimals to two decimal places.

NS 3.0, **3.3** Solve problems involving multiplication of multidigit numbers by two-digit number.

MR 3.0, 3.3 Develop generalizations of the results obtained and apply them in other circumstances.

• Multiplying Multiples of 10 and 100, Part 2

Power Up

facts   Power Up G

mental math

Use the fives pattern as you add in problems **a–c.**

a. Number Sense: 36 + 15

b. Number Sense: 47 + 25

c. Number Sense: 28 + 35

d. Number Sense: 40 × 40 × 10

e. Money: $10.00 − $2.75

f. Time: How many days is 8 weeks?

g. Estimation: Each bracelet costs $2.99. Kim has $11. Does she have enough money to buy 4 bracelets?

h. Calculation: $\frac{1}{2}$ of 42, ÷ 3, + 10, − 3, ÷ 2, × 7

problem solving

The diagram at right is called a *Venn diagram.* The circle on the left represents fruit, and the circle on the right represents vegetables. The *A* represents apples, which are a fruit, and the *B* represents broccoli, which is a vegetable. The *C* represents cheese, which is neither a fruit nor a vegetable. Copy the diagram on your paper and place abbreviations for eggs, oranges, and green beans.

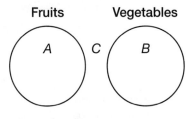

New Concept

Thinking Skills

Analyze

Is the product of 40 and 50 written as 200, 2000, or 20,000? Explain your reasoning.

Once we have memorized the multiplication facts, we can multiply rounded numbers "in our head." To do this, we multiply the first digits of the factors and count zeros. Study the multiplication on the next page.

$$\begin{array}{r} 40 \\ \times\ 30 \\ \hline \underline{} \end{array}$$

40 ← two zeros
× 30 ←

4 × 3 → two zeros

To find the product of 40 and 30, we multiply 4 by 3 and then attach two zeros.

Example 1

In the weightlifting room, a group of football players lifted 80 pounds of weights 60 different times. How many pounds of weight did the players lift altogether?

We think, "six times eight is 48." Since there is one zero in 60 and one zero in 80, we attach two zeros to 48. The product is 4800, so the total weight lifted was **4800 pounds.**

6 × 8 is 48.

Example 2

Thinking Skills

Verify

Why do we attach three zeros when we multiply 30 by $7.00?

A store has 30 ping pong paddles for sale at $7.00 each. How much money will the store receive if all of the paddles are sold?

We think, "three times seven is 21." There are three zeros in the problem, so we attach three zeros to 21 to get 21,000. Since we multiplied dollars and cents, we insert the decimal point two places from the right and add a dollar sign. The product is $210.00, so the income will be **$210.**

3 × 7 is 21.

Example 3

Multiply mentally: 400 × 700

We think, "Four times seven is 28." We ○ ○ ○ attach four zeros and get **280,000**.

(4 × 7 is 28.)

Connect How would we multiply 40 × 7000? What is the product?

Lesson Practice Multiply mentally:

a. 70 × 80 **b.** 40 × 50

c. 40 × $6.00 **d.** 30 × 800

e. **Verify** Write >, <, or = to make this statement true:
300 × 200 ◯ 30 × 2000

Written Practice *Distributed and Integrated*

1. **Analyze** Three quarters, four dimes, two nickels, and seven pennies is how much money?
(28)

*2. **Formulate** Write a division word problem with a quotient of 630.
(Inv. 1, 79)

*3. **Explain** Gregory paid $1 for a folder and received 52¢ in change. If the tax was 3¢, how much did the folder cost without tax? Explain your thinking.
(76)

4. Ryan wrote each of his 12 spelling words five times. In all, how many words did he write?
(39)

5. **Estimate** In the 2004 presidential election, 5992 voters in Blaine County, Idaho, voted for candidate John Kerry, and 4034 voters voted for candidate George Bush. Estimate the total number of votes those two candidates received, and explain your estimate.
(40)

6. What is the tally for 10?
(Inv. 6)

7. Name the shaded part of this square
(Inv. 4)
   a. as a fraction.

   b. as a decimal number.

8. **Represent** One sixth of the 48 crayons are in the box. How many
(74) crayons are in the box? Draw a picture to illustrate the problem.

9. Segment *AB* is 32 mm long. Segment *BC* is 26 mm long. Segment
(33, 42) *AD* is 91 mm long. How many millimeters long is segment *CD*?

          A   *B*   *C*   *D*

* **10.** Which digit in 6.120 is in the hundredths place?
(41)

11. **Estimate** If a pint of water weighs about one pound, then about how
(73, 80) many pounds does a quart of water weigh?

* **12.** $4.32 - 0.43$
(45)

13. $5^2 + \sqrt{25} + n = 30$
(Inv. 3, 63, 65)

14. $\$6.08$
(59) $\times$ 8

* **15.** 47
(71) $\times\ 20$

* **16.** 300
(85) $\times\ \ 20$

17. 53×30
(71)

* **18.** 63×40
(71)

19. 100×32
(84)

20. $4\overline{)3456}$
(79)

21. $8n = 6912$
(79)

22. $7\overline{)\$50.40}$
(79, 83)

* **23.** Draw a right, scalene triangle.
(81)

24. **Represent** **a.** Draw a square with sides 4 cm long.
(18, 19
Inv. 3)

    b. Shade half of the square you drew. How many square centimeters
       did you shade?

* **25.** **Represent** Write twenty-one hundredths as a fraction and as a
(Inv. 4) decimal number.

* 26.
(53, 72)
Explain Emma mixed two quarts of orange juice from frozen concentrate. She knows 1 quart is equal to 32 fluid ounces. The small juice glasses Emma is filling each have a capacity of 6 fluid ounces. How many juice glasses can Emma fill? Explain your answer.

* 27. Multiple Choice Use the polygons below to answer parts **a–d.**
(17, 82)

   a. Which of these polygons has no lines of symmetry?

   b. Which two of these polygons have rotational symmetry?

   c. Which polygon does not have any parallel sides?

   d. Which polygons do not have any perpendicular sides?

28. How many degrees does the minute hand of a clock turn in half
(78) an hour?

* 29. Compare: 4.2 ◯ 4.200
(44)

* 30. Use the pictograph below to answer the questions that follow:
(Inv. 5)

Animal	Typical Weight (in pounds)
Alligator	⊙—⊙ ⊙—
Porpoise	⊙—⊙
Wild Boar	⊙—⊙ ⊙—⊙ ⊙—⊙
Seal	⊙—⊙ ⊙—⊙

Key: ⊙—⊙ = 100 pounds

   a. What amount of weight does each symbol represent?

   b. Write the typical weights of the animals in order from least to greatest.

   c. **Connect** Write a sentence that compares the weights of two animals.

86

California Mathematics Content Standards

SDAP 2.0, 2.1 Represent all possible outcomes for a simple probability situation in an organized way (e.g., tables, grids, tree diagrams)

MR 2.0, 2.4 Express the solution clearly and logically by using the appropriate mathematical notation and terms and clear language; support solutions with evidence in both verbal and symbolic work.

• Multiplying Two Two-Digit Numbers, Part 1

Power Up

facts

Power Up G

mental math

Use the fives pattern as you subtract in problems **a–c.**

a. Number Sense: $41 - 15$

b. Number Sense: $72 - 25$

c. Number Sense: $84 - 45$

d. Number Sense: 25×30

e. Money: Bridget spent $6.54. Then she spent $2.99 more. Altogether, how much did Bridget spend?

f. Time: Mirabel's speech lasted 2 minutes 20 seconds. How many seconds is that?

g. Estimation: Kione purchased two DVDs for $18.88 each. Estimate the total cost of the DVDs.

h. Calculation: $\frac{1}{10}$ of 60, $\times$ 4, $\div$ 2, $\times$ 5

problem solving

Choose an appropriate problem-solving strategy to solve this problem. Josh will flip a coin three times in a row. On each flip, the coin will either land on "heads" or "tails." If the coin were to land heads up each time, the combination of flips would be heads, heads, heads, which can be abbreviated as HHH. Find all the possible combinations of heads and tails Josh can get with three coin flips.

New Concept

We use three steps to multiply by a two-digit number. First we multiply by the ones digit. Next we multiply by the tens digit. Then we add the products. To multiply 34 by 12, for example, we multiply 34 by 2 and then multiply 34 by 10. Then we add the products.

572   *Saxon Math Intermediate 4*

$$34 \times 2 = 68 \quad \text{partial product}$$
$$34 \times 10 = 340 \quad \text{partial product}$$
$$34 \times 12 = 408 \quad \text{total product}$$

Thinking Skills

Evaluate

Can the expression $(2 \times 34) + (10 \times 34)$ be used to represent the vertical form of 34×12? Explain why or why not.

It is easier to write the numbers one above the other when we multiply, like this:

$$\begin{array}{r} 34 \\ \times\ 12 \end{array}$$

Method 1: First we multiply 34 by 2 and write the answer.

$$\begin{array}{r} 34 \\ \times\ 12 \\ \hline 68 \end{array}$$

Next we multiply 34 by 1. This 1 is actually 10, so the product is 340. We write the answer, and then we add the results of the two multiplication problems and get 408.

$$\begin{array}{r} 34 \\ \times\ 12 \\ \hline 68 \\ 340 \\ \hline 408 \end{array}$$

Method 2: An alternate method would be to omit the zero from the second multiplication. Using this method, we position the last digit of the second multiplication in the second column from the right. The empty place is treated like a zero when adding.

$$\begin{array}{r} 34 \\ \times\ 12 \\ \hline 68 \\ 34 \\ \hline 408 \end{array}$$

Example

Multiply:
$$\begin{array}{r} 31 \\ \times\ 23 \end{array}$$

First we multiply 31 by 3.

$$\begin{array}{r} 31 \\ \times\ 23 \\ \hline 93 \end{array}$$

Now we multiply 31 by 2. Since this 2 is actually 20, we write the last digit of the product in the tens column. Then we add to get **713.**

$$
\begin{array}{r}
31 \\
\times\ 23 \\
\hline
93 \\
62 \\
\hline
713
\end{array}
\quad \text{or} \quad
\begin{array}{r}
31 \\
\times\ 23 \\
\hline
93 \\
620 \\
\hline
713
\end{array}
$$

Lesson Practice   Multiply:

a. $\begin{array}{r} 32 \\ \times\ 23 \\ \hline \end{array}$ **b.** $\begin{array}{r} 25 \\ \times\ 32 \\ \hline \end{array}$ **c.** $\begin{array}{r} 43 \\ \times\ 12 \\ \hline \end{array}$ **d.** $\begin{array}{r} 34 \\ \times\ 21 \\ \hline \end{array}$

e. $\begin{array}{r} 32 \\ \times\ 32 \\ \hline \end{array}$ **f.** $\begin{array}{r} 22 \\ \times\ 14 \\ \hline \end{array}$ **g.** $\begin{array}{r} 13 \\ \times\ 32 \\ \hline \end{array}$ **h.** $\begin{array}{r} 33 \\ \times\ 33 \\ \hline \end{array}$

Written Practice   *Distributed and Integrated*

*1. **(Analyze)** Use this information to answer parts **a–c.**
(76)

   Maritza invited 2 friends over for lunch. She plans to make 2 tuna sandwiches, a bologna sandwich, and 3 chicken sandwiches.

   a. How many sandwiches will Maritza make in all?

   b. Including Maritza, each person can have how many sandwiches?

   c. If Maritza cuts each tuna sandwich in half, how many halves will there be?

2. Five pounds of grapes cost $2.95. What is the cost per pound?
(53, 79)

*3. If each side of a hexagon is 4 inches long, what is the perimeter of the
(20, 42, 50) hexagon in feet?

4. **(Represent)** Four hundred fifty-seven thousand is how much greater
(27, 52) than three hundred eighty-four thousand, nine hundred seventy-six?

*5. Three brands of whole-grain cereal cost $4.68, $4.49, and $4.71.
(Inv. 4) Arrange these prices in order from least to greatest.

*6. **(Estimate)** Lauren saw that a spindle of 50 blank CDs costs $9.79. Estimate the cost for Lauren to buy 100 blank CDs. Explain your answer.
(61, 62)

7. Name the shaded part of the large square
(Inv. 4)
    a. as a fraction.

    b. as a decimal number.

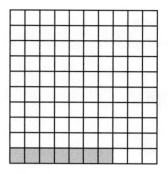

8. **(Represent)** Use words to write $7572\frac{1}{8}$.
(32)

*9. **(Represent)** At Kelvin's school, one fifth of the 80 fourth grade students ride the bus to and from school each day. How many fourth grade students ride the bus? Draw a picture to illustrate the problem.
(74)

*10. **(Analyze)** Josh has $46 in his bank account. He wrote a check for $53. What number represents his account balance?
(76)

11. Franca's trip only lasted for a couple of hours. According to the clocks shown below, exactly how long did the trip take?
(13)

Began   Finished

*12. **(Justify)** James traveled 301 miles in 7 hours. He traveled an average of how many miles per hour? Explain your reasoning.
(60)

*13. Martino bought 3 folders priced at $1.99 each. Sales tax was 33¢. He paid with a $20 bill. How much money should he get back?
(51, 52, 59)

14. $25 + $2.75 + $15.44 + 27¢
(28)

* 15. $m + 0.26 = 6.2$
(8, 45)

16. $\$100 - \$89.85$
(45, 52)

17. 65×1000
(84)

18. 42×30
(71)

19. 21×17
(86)

* 20. 368
(59) $\times \quad 4$

* 21. 4000
(85) $\times \quad 20$

22. $\$4.79$
(59) $\times \quad 6$

23. $9\overline{)918}$
(83)

24. $5r = 485$
(34, 69)

25. $6\overline{)482}$
(83)

26. $\$50.00 \div 8$
(79)

27. $2100 \div 7$
(83)

28. $0.875 - (0.5 + 0.375)$
(9, 45)

* 29. a. (Verify) Which of these letters has two lines of symmetry?
(82)

H A P P Y

b. Which of these letters has rotational symmetry?

* 30. (Represent) Draw a triangle that has two perpendicular sides. What
(81) type of triangle did you draw?

California Mathematics Content Standards

NS **3.0, 3.4** Solve problems involving division of multidigit numbers by one-digit numbers.

MR 1.0, 1.1 Analyze problems by identifying relationships, distinguishing relevant from irrelevant information, sequencing and prioritizing information, and observing patterns.

• Remainders in Word Problems

facts   Power Up G

mental math   Use the fives pattern as you add or subtract in problems **a–c**.

    a. Number Sense: $83 - 15$

    b. Number Sense: $29 + 35$

    c. Number Sense: $76 + 15$

    d. Fractional Part: Corey figures that about $\frac{1}{2}$ of the calories he consumes are from carbohydrates. Corey consumes about 2000 calories each day. About how many of those calories are from carbohydrates?

    e. Measurement: How many inches is one yard?

    f. Time: Which day of the week is 71 days after Monday?

    g. Estimation: Jayla has run $\frac{1}{2}$ mile in 4 minutes 57 seconds. If she can continue running at the same pace, about how long will it take Jayla to run one full mile?

    h. Calculation: $5^2 + 5^2, + 6, \div 8$

problem solving   Choose an appropriate problem-solving strategy to solve this problem. Sandra bought a CD priced at $12.95. **Sales tax** was $1.10. She paid for her purchase with a $10 bill and a $5 bill. Sandra got back five coins (not including a half-dollar). What were the coins Sandra should have received in change?

New Concept

We have practiced solving "equal groups" problems using division. In these problems, there were no remainders from the division. In this lesson we will begin practicing division word problems that involve remainders. When solving these problems, we must be careful to identify exactly what the question is asking.

The packer needs to place 100 bottles into boxes that hold 6 bottles each.

   a. How many boxes can be filled?

   b. How many bottles will be left over?

   c. How many boxes are needed to hold all the bottles?

Each of these questions asks for different information. To answer the questions, we begin by dividing 100 by 6.

$$\begin{array}{r} 16 \text{ R } 4 \\ 6\overline{)100} \\ \underline{6} \\ 40 \\ \underline{36} \\ 4 \end{array}$$

The result "16 R 4" means that the 100 bottles can be separated into 16 groups of 6 bottles. There will be 4 extra bottles.

   a. The bottles can be separated into 16 groups of 6 bottles, so **16 boxes** can be filled.

   b. The 4 remaining bottles do not completely fill a box. So after filling 16 boxes, there will still be **4 bottles** left over.

   c. Although the 4 remaining bottles do not completely fill a box, another box is needed to hold them. Thus, **17 boxes** are needed to hold all the bottles.

Lesson Practice

Interpret Use the statements below to answer problems **a–e.**

Tomorrow 32 students are going on a field trip. Each car can carry 5 students.

   a. How many cars can be filled?

   b. How many cars will be needed?

Tendai found 31 quarters in his bank. He made stacks of 4 quarters each.

   c. How many stacks of 4 quarters did he make?

   d. How many extra quarters did he have?

   e. If Tendai made a short stack with the extra quarters, how many stacks would he have in all?

* 1. Ninety-one students are divided as equally as possible among
(87) 3 classrooms.

    a. How many classrooms have exactly 30 students?

    b. How many classrooms have 31 students?

2. (**Analyze**) **a.** 1970 it cost 6¢ to mail a letter. How much did it cost to
(39) mail twenty letters in 1970?

    b. How much does it cost to mail twenty letters today?

3. 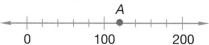 (**Represent**) Point *A* represents what number on this number
(Inv. 2) line?

$$\begin{array}{c} A \\ \longleftarrow\!\!+\!\!+\!\!+\!\!+\!\!+\!\!+\!\!\bullet\!\!+\!\!+\!\!+\!\!+\!\!\longrightarrow \\ 0 \qquad 100 \qquad 200 \end{array}$$

4. George Washington was president of the United States of America until
(52) 1796. How many years has it been since his presidency?

5. A $1 bill weighs about 1 gram. How much would a $5 bill weigh?
(80)

6. Name the shaded part of the large square
(Inv. 4)

    a. as a fraction.

    b. as a decimal number.

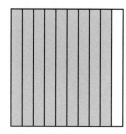

* 7. (**Evaluate**) If $y = 4x + 5$, then what is y when $x = 4, 5,$ and 6?
(64,
Inv. 7)

* 8. a. A regular pentagon has how many lines of symmetry?
(82)

    b. (**Justify**) Does a regular pentagon have rotational symmetry? How
do you know?

* 9. (**Represent**) One half of the 32 chess pieces were still on the board.
(74) How many chess pieces were still on the board? Draw a picture to
illustrate the problem.

10. Miriam left home at 10:30 a.m. She traveled for 7 hours. What time was
(13) it when she arrived?

11. Maureo traveled 42 miles in 1 hour. If he kept going at the same speed,
(58, 71) how far would he travel in 20 hours?

* **12.** Violet gave the cashier $40 for a toaster that cost $29.99 plus $1.80
(52) in tax. What was her change? Write an equation to solve the problem.

* **13.** Alvin faced the sun as it set in the west, then turned 90°
(78) counterclockwise and headed home. In what direction was Alvin
heading after he turned?

14. $n + 8 + 2 + 3 + 5 + 2 = 24$
(4)

15. $4.12 - (3.6 + 0.2 + 0.12)$
(9, 45)

16. $18 - 15.63
(28, 52)

17. $15.27 + 85.75
(28, 51)

18. $2^3 \times \sqrt{25}$
(Inv. 3, 65)

19. 30×90
(85)

20. 7.50×8
(59)

* **21.**   14
(86) $\times\ 22$

* **22.**   5126
(59) $\times\quad 4$

23.   74
(71) $\times\ 40$

24. $4\overline{)\$6.36}$
(79)

25. $5\overline{)800}$
(83)

26. $473 \div 8$
(72)

27. $3m = 1800$
(34, 83)

* **28.** 16×1000
(84)

* **29.** 263×100
(84)

30. Find the perimeter and area of this rectangle.
(66)

50 ft

20 ft

California Mathematics Content Standards

NS 1.0, **1.3** Round whole numbers through the millions to the nearest ten, hundred, thousand, ten thousand, or hundred thousand.

MR 2.0, 2.6 Make precise calculations and check the validity of the results from the context of the problem.

MR 3.0, 3.1 Evaluate the reasonableness of the solution in the context of the original situation.

• Multiplying Two Two-Digit Numbers, Part 2

Power Up

facts　　Power Up G

mental math

　a. Number Sense: $85 - 38$

　b. Number Sense: $4 \times 20 \times 10$

　c. Fractional Part: $\frac{1}{10}$ of $20

　d. Measurement: How many pints is a gallon?

　e. Powers/Roots: $9^2 - \sqrt{9}$

　f. Time: Which day of the week is 699 days after Monday?

　g. Estimation: Estimate the fraction of this circle that is shaded.

　h. Calculation: $\sqrt{81}$, $\div\ 3$, $\times\ 25$, $+\ 75$, $\times\ 2$

problem solving

Choose an appropriate problem-solving strategy to solve this problem. Tandy wants to know the circumference of (the distance around) her bicycle tire. She has some string and a meterstick. How can Tandy measure the circumference of the tire in centimeters?

New Concept

Recall the three steps for multiplying two two-digit numbers:

Step 1: Multiply by the ones digit.

Step 2: Multiply by the tens digit.

Step 3: Add to find the total.

Example 1

A college auditorium has 27 rows of seats and 46 seats in each row. How many people can be seated in the auditorium?

The first step is to multiply 46 by 7. The result is 322. This is not the final product. It is called a partial product.

$$
\text{Step 1} \quad
\begin{array}{r}
4 \\
46 \\
\times\ 27 \\
\hline
322
\end{array}
$$

The second step is to multiply 46 by the 2 of 27. Since we are actually multiplying by 20, we place a zero in the ones place or shift this partial product one place to the left.

$$
\text{Step 2} \quad
\begin{array}{r}
1 \\
4 \\
46 \\
\times\ 27 \\
\hline
322 \\
92 \\
\hline
\end{array}
\quad \text{or} \quad
\begin{array}{r}
1 \\
4 \\
46 \\
\times\ 27 \\
\hline
322 \\
920 \\
\hline
\end{array}
$$

$$\text{Step 3} \quad 1242 \qquad\qquad 1242$$

The third step is to add the partial products. The final product is 1242.

We find that **1242 people** can be seated.

Example 2

A golf course has 46 different viewer mounds. Each mound can seat an average of 72 viewers. How many viewers in all can be seated on the mounds?

First we multiply 46 by 2.

$$
\begin{array}{r}
1 \\
46 \\
\times\ 72 \\
\hline
92
\end{array}
$$

Next we multiply 46 by 7 and then add the partial products.

$$
\begin{array}{r}
4 \\
1 \\
46 \\
\times\ 72 \\
\hline
92 \\
322 \\
\hline
3312
\end{array}
\quad \text{or} \quad
\begin{array}{r}
4 \\
1 \\
46 \\
\times\ 72 \\
\hline
92 \\
3220 \\
\hline
3312
\end{array}
$$

We find that **3312 viewers** can be seated.

Example 3

Adelio estimated the product of 86 × 74 to be 6300. Did Adelio make a reasonable estimate?

Before multiplying, we round 86 to 90 and round 74 to 70. Since 90 × 70 = 6300, Adelio's estimate is reasonable. (The exact product is 6364.)

Lesson Practice   Multiply:

a.	38	**b.**	49	**c.**	84	**d.**	65
	× 26		× 82		× 67		× 48

e. Mya is renting 21 tables for a reception. The rental charge is $29 per table. Explain how Mya can make a reasonable estimate of the total cost.

Written Practice   *Distributed and Integrated*

1. Joel gave the clerk a $5 bill to pay for a half gallon of milk that cost $1.06
(28, 51, 52) and a box of cereal that cost $2.39. How much change should he receive?

*2. What fraction of the letters in the following word are A's?
(77)
ALABAMA

3. Melba planted 8 rows of apple trees. There were 15 trees in each row.
(39) How many trees did she plant?

4. A ruble is a Russian coin. If four pounds of beets costs one hundred fifty-six rubles,
(53, 69) what is the cost in rubles of each pound of beets?

*5. a. This scale shows a mass of how many grams?
(80)

b. **Explain** Would this fruit have the same mass on another planet? Explain why.

6. Name the shaded part of the large square
(Inv. 4)

    a. as a fraction.

    b. as a decimal number.

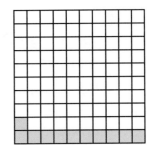

* **7.** (Interpret) Peter packed 6 table-tennis balls in each package. There
(87) were 100 table-tennis balls to pack.

    a. How many packages did he fill?

    b. How many table-tennis balls were left over?

* **8.** (List) Write the factors of 35.
(55)

9. (Represent) Bactrian camels have 2 humps. One third of the
(74) 24 camels were Bactrian. How many camels were Bactrian? Draw
  a picture to illustrate the problem.

10. A quart is a quarter of a gallon. A quart is what decimal part of a gallon?
(Inv. 4,
73)

* **11.** (Classify) For each statement, write either *true* or *false*:
(81)

    a. Every right triangle has perpendicular sides.

    b. Every isosceles triangle is also an equilateral triangle.

* **12. a.** (Represent) Seventy-one of the one hundred students in the
(Inv. 4) school were girls. Girls made up what fraction of the students in
    the school?

    b. (Represent) Write your answer for part **a** as a decimal number.
    Then use words to name the number.

* **13.** Which digit in 1.875 is in the tenths place?
(41)

* **14.** If $y = 2x - 3$, what is y when x is 5?
(64)

15. Tyler traveled 496 miles in 8 hours. He traveled an average of how
(60, 68) many miles per hour?

* 16. Find $8.3 - (1.74 + 0.9)$. Describe the steps in order.
(9, 45)

17. 63×1000
(84)

18. $80 \times 50¢$
(85)

19. $\begin{array}{r} 37 \\ 81 \\ 45 \\ 139 \\ 7 \\ 15 \\ + \ 60 \\ \hline \end{array}$
(11)

* 20. $\begin{array}{r} 52 \\ \times \ 15 \\ \hline \end{array}$
(88)

* 21. $\begin{array}{r} 36 \\ \times \ 27 \\ \hline \end{array}$
(88)

22. $2\overline{)714}$
(79)

23. $6\overline{)789}$
(79)

24. $3n = 624$
(34, 79)

25. $5 + w = 5^2$
(64, 65)

* 26. (Model) Write five ordered pairs for the equation $y = 2x$. Then use
(Inv. 7) **Lesson Activity 8** to graph the ordered pairs.

* 27. (Analyze) Write these numbers in order from least to greatest:
(43)
$$\frac{1}{10}, \ 2.06, \ 1\frac{4}{10}, \ \frac{3}{100}$$

28. A room is 5 yards long and 4 yards wide. How many square yards of
(Inv. 3) carpeting are needed to cover the floor?

29. The radius of this circle is 15 millimeters. The diameter of the
(18, 42) circle is how many centimeters?

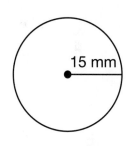

15 mm

* 30. a. (Verify) Which of these letters has two lines of symmetry?
(17, 82)

V W X Y Z

b. (Verify) Which two letters have rotational symmetry?

c. Multiple Choice The angle formed by the letter V illustrates what
kind of angle?

    A acute   **B** right   **C** obtuse   **D** straight

LESSON
89

• Mixed Numbers and Improper Fractions

California Mathematics Content Standards

NS 1.0, 1.5 Explain different interpretations of fractions, for example, parts of a whole, parts of a set, and division of whole numbers by whole numbers; explain equivalents of fractions (see Standard 4.0)

NS 1.0, 1.7 Write the fraction represented by a drawing of parts of a figure; represent a given fraction by using drawings; and relate a fraction to a simple decimal on a number line.

facts   Power Up G

mental math

    a. Number Sense: 25×1000

    b. Number Sense: $58 + 35$

    c. Fractional Part: Alonso needs to collect $\frac{1}{4}$ of $40. What is $\frac{1}{4}$ of $40?

    d. Time: What day is 71 days after Wednesday?

    e. Measurement: How many feet is 6 yards?

    f. Money: The book cost $6.75. If Daina paid for the book with a $10 bill, then how much change should she receive?

    g. Estimation: The total cost for 6 picture frames was $41.94. Round this amount to the nearest dollar and then divide by 6 to estimate the cost of each frame.

    h. Calculation: $\sqrt{1}, \times 1, \div 1, -1 + 1$

problem solving

In this Venn diagram, the circle on the left represents multiples of 3. The circle on the right represents even numbers. The number 6 is both a multiple of 3 and an even number, so it is placed within the space created by the overlap of the two circles. The number 4 is placed within the circle for even numbers but outside the overlap, since 4 is not a multiple of 3. The number 1 is placed outside both circles because it is not a multiple of 3 and it is not even. Copy the Venn diagram on your paper, and place the numbers 9, 10, 11, and 12.

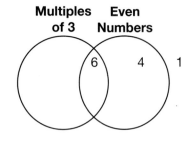

Here we show a picture of $1\frac{1}{2}$ shaded circles. Each whole circle has been divided into two half circles.

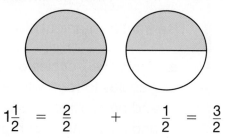

$$1\frac{1}{2} \;=\; \frac{2}{2} \quad + \quad \frac{1}{2} \;=\; \frac{3}{2}$$

Math Language

A **proper fraction** is a fraction whose numerator is less than the denominator.

We see from the picture that $1\frac{1}{2}$ is the same as three halves, which is written as $\frac{3}{2}$. The numerator is greater than the denominator, so the fraction $\frac{3}{2}$ is greater than 1. Fractions that are greater than or equal to 1 are called **improper fractions.** In this lesson we will draw pictures to show mixed numbers and their equivalent improper fractions.

Example 1

Draw circles to show that $2\frac{3}{4}$ equals $\frac{11}{4}$.

We begin by drawing three circles. The denominator of the fraction part of $2\frac{3}{4}$ is four, so we divide all the circles into fourths and shade $2\frac{3}{4}$ of them.

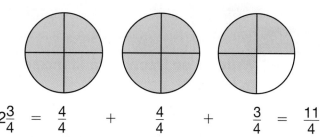

$$2\frac{3}{4} \;=\; \frac{4}{4} \quad + \quad \frac{4}{4} \quad + \quad \frac{3}{4} \;=\; \frac{11}{4}$$

We count 11 shaded fourths. The drawing shows that $2\frac{3}{4}$ equals $\frac{11}{4}$.

Activity

Modeling Mixed Numbers and Improper Fractions

Material needed:

- fraction manipulatives from **Lesson Activities 11** and **12**

Use fraction manipulatives to perform the following activities:

a. Place five $\frac{1}{2}$ circles on a desk. Then arrange four of the $\frac{1}{2}$ circles to form whole circles. Draw a picture of the whole and part circles you formed, and write the improper fraction and mixed number represented by the five $\frac{1}{2}$ circles.

b. Place one more $\frac{1}{2}$ circle on the desk to complete another circle. Write the improper fraction and whole number represented.

c. Clear the desk of $\frac{1}{2}$ circles and place seven $\frac{1}{4}$ circles on the desk. Fit the pieces together to form a whole circle and part of a circle. Draw a picture, and write the improper fraction and mixed number represented.

d. Place one more $\frac{1}{4}$ circle on the desk to complete another circle. Write the improper fraction and whole number represented.

Example 2

Write these fractions and mixed numbers in order from greatest to least.

$$1\frac{2}{5}, \frac{4}{5}, 1\frac{3}{4}$$

We start by writing the mixed numbers as fractions.

$$1\frac{2}{5} = \frac{5}{5} + \frac{2}{5} \text{ or } \frac{7}{5}$$

We have two fractions that are fifths: $\frac{7}{5}$ and $\frac{4}{5}$. We can see that $\frac{7}{5}$ is greater than $\frac{4}{5}$.

We write $\frac{7}{5}, \frac{4}{5}$.

Now we can write $1\frac{3}{4}$ as a fraction.

$$1\frac{3}{4} = \frac{4}{4} + \frac{3}{4} \text{ or } \frac{7}{4}$$

Finally we compare fifths and fourths.
Since $\frac{1}{4}$ is greater than $\frac{1}{5}$, then $\frac{7}{4}$ is greater than $\frac{7}{5}$.
We write $\frac{7}{4}, \frac{7}{5}, \frac{4}{5}$.

Discuss Why did we use $\frac{5}{5}$ to change $1\frac{2}{5}$ to $\frac{7}{5}$?

Lesson Practice

a. Draw circles to show that $1\frac{3}{4} = \frac{7}{4}$.

b. Draw circles to show that $2\frac{1}{2} = \frac{5}{2}$.

c. Draw circles to show that $1\frac{1}{3} = \frac{4}{3}$.

d. Write these numbers from greatest to least.

$$\frac{5}{2}, \frac{3}{4}, 1\frac{1}{4}$$

Written Practice Distributed and Integrated

1. **Interpret** Use this tally sheet to answer parts **a–c**.
(Inv. 6)

Results of Class Election

Candidate	Tally
Irma	₌卄 II
Hamish	卄 I
Thanh	卄 III
Marisol	卄 卄 II

a. Who was second in the election?

b. Who received twice as many votes as Hamish?

c. Altogether, how many votes were cast?

2. Write these amounts in order from greatest to least:
(Inv. 4)
$1.45 $2.03 $0.99 $1.48

3. **Formulate** The Osage River in Kansas is 500 miles long. The
(14, 16) Kentucky River is 259 miles long. How many miles longer is the Osage River? Write and solve an equation.

*4. What fraction of the letters in the following word are I's?
(77)

MISSISSIPPI

*5. (**Analyze**) The coach divided 33 players as equally as possible into
(87) 4 teams.

    a. How many teams had exactly 8 players?

    b. How many teams had 9 players?

*6. (**Conclude**) Write *true* or *false* for parts **a–d.**
(81)
    a. All equilateral triangles are also equiangular.

    b. A right triangle could have more than one right angle.

    c. An equilateral triangle is also an isosceles triangle.

    d. All triangles are polygons.

7. Name the shaded part of this group
(Inv. 4,
77)   **a.** as a fraction.

    b. as a decimal number.

*8. Estimate the product of 88 and 59. Then find the exact product.
(88)

9. Sue's birthday is May 2. Her birthday will be on what day
(RF12) of the week in the year 2045?

MAY 2045						
S	M	T	W	T	F	S
	1	2	3	4	5	6
7	8	9	10	11	12	13
14	15	16	17	18	19	20
21	22	23	24	25	26	27
28	29	30	31			

*10. Point *W* represents what number on this number line?
(Inv. 2)

11. $32.63 + $42 + $7.56
(28, 51)

12. $86.45 − ($74.50 + $5)
(9, 51, 52)

13. 83×40
(71)

14. 1000×53
(84)

15. 200×800
(85)

*__16.__
(88)
$$\begin{array}{r} 32 \\ \times\ 16 \\ \hline \end{array}$$

*__17.__
(88)
$$\begin{array}{r} 67 \\ \times\ 32 \\ \hline \end{array}$$

18.
(59)
$$\begin{array}{r} \$8.95 \\ \times\quad 4 \\ \hline \end{array}$$

19. $3\overline{)625}$
(83)

20. $4\overline{)714}$
(79)

21. $6\overline{)1385}$
(83)

22. $\dfrac{900}{5}$
(83)

23. $3748 \div 9$
(79)

24. $8m = \$28.56$
(34, 79)

*__25.__ **Represent** This circle shows that $\frac{2}{2}$ equals 1. Draw a circle that
(89) shows that $\frac{3}{3}$ equals 1.

26. Find the perimeter and area of this rectangle.
(66)

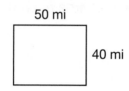

50 mi

40 mi

*__27.__ **a.** Draw two congruent isosceles triangles.
(70, 81, 82)

   b. Draw the line of symmetry on one of the figures you created.

*__28.__ Compare: $0.5 \bigcirc 0.50$
(44)

*__29.__ Kelly ran and jumped 9 ft 6 in. How many inches did Kelly jump?
(42)

* 30. The table shows the relationship between the number of hours Aidan
(58,
Inv. 7) works and the amount of money he earns.

Number of Hours Worked	Income Earned (in dollars)
1	19
2	38
3	57
4	76
5	95

a. **Generalize** Write a word sentence that describes the relationship of the data.

b. **Predict** Aidan works 40 hours each week. What is a reasonable estimate of the amount of income he earns each week? Explain your answer.

• Classifying Quadrilaterals

California Mathematics Content Standards

MG 3.0, 3.1 Identify lines that are parallel and perpendicular.

MG 3.0, 3.5 Know the definitions of a right angle, an acute angle, and an obtuse angle. Understand that 90°, 180°, 270°, and 360° are associated, respectively, with $\frac{1}{4}$, $\frac{1}{2}$, $\frac{3}{4}$, and full turns.

MG 3.0, 3.8 Know the definition of different quadrilaterals (e.g., rhombus, square, rectangle, parallelogram, trapezoid).

facts   Power Up I

mental math   Find half of each number in problems **a–d.**

    a. Number Sense: 40

    b. Number Sense: 48

    c. Number Sense: 64

    d. Number Sense: 86

    e. Number Sense: 75 + 37

    f. Money: Taylor bought scissors for $3.54 and glue for $2.99. What was the total cost?

    g. Estimation: Choose the more reasonable estimate for the mass of 500 sheets of copy paper: 2 grams or 2 kilograms.

    h. Calculation: $\sqrt{49}$, × 2, + 7, ÷ 3, × 7

problem solving   Choose an appropriate problem-solving strategy to solve this problem. A half-ton pickup truck can carry a load weighing half of a ton. How many 100-pound sacks of cement can a half-ton pickup truck carry?

New Concept

Recall from Lesson 50 that a quadrilateral is a polygon with four sides. In this lesson we will practice recognizing and naming different types of quadrilaterals. On the following page, we show four different types.

A   B   C   D   E

parallelogram   parallelogram   parallelogram   parallelogram   trapezoid
                 rhombus   rectangle   rhombus
                                                   rectangle
                                                   square

A **parallelogram** is a quadrilateral with *two* pairs of parallel sides. Figures *A, B, C,* and *D* each have two pairs of parallel sides, so all four figures are parallelograms. A **trapezoid** is a quadrilateral with exactly *one* pair of parallel sides. Figure *E* is not a parallelogram; it is a trapezoid.

A **rectangle** is a special type of parallelogram that has four right angles. Figures *C* and *D* are rectangles. A **rhombus** is a special type of parallelogram whose sides are equal in length. Figure *B* is a rhombus, as is figure *D*. A **square** is a regular quadrilateral. Its sides are equal in length, and its angles are all right angles. Figure *D* is a square. It is also a parallelogram, a rhombus, and a rectangle.

Example 1

Which of these quadrilaterals is *not* a parallelogram?

F   G   H   I

We look for pairs of parallel sides. A parallelogram has two pairs of parallel sides. Figures *F, G,* and *I* each have two pairs of parallel sides. **Figure H** has only one pair of parallel sides, so it is a trapezoid, not a parallelogram.

Example 2

Draw two parallel line segments of different lengths. Then form a quadrilateral by drawing two line segments that connect the endpoints. What type of quadrilateral did you make?

First we draw two parallel line segments of different lengths.

Then we connect the endpoints with line segments to form a quadrilateral.

We see that this quadrilateral is a **trapezoid.**

Example 3

Thinking Skills

Model

Find a quadrilateral in your classroom. Identify and describe the parallel, perpendicular, and intersecting segments in the quadrilateral.

Which of the following quadrilaterals has sides that are *not* parallel or perpendicular?

A

B

C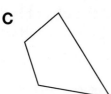

D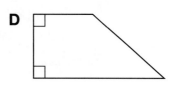

We will consider the relationships between the sides of each quadrilateral.

A The opposite sides are parallel, and the adjacent sides are perpendicular.

B The opposite sides are parallel, and the adjacent sides intersect but are not perpendicular.

C There are no parallel or perpendicular sides.

D One pair of opposite sides is parallel, and another side is perpendicular to the parallel sides.

Only **figure C** has sides that are not parallel or perpendicular.

Example 4

Describe the angles in each of the quadrilaterals in Example 3.

Figure **A** appears to be a square; it has **four right angles.**

Figure **B** is a parallelogram; it has **two acute angles and two obtuse angles.**

Figure **C** is a quadrilateral with **two obtuse angles and two acute angles.**

Figure **D** is a trapezoid with **two right angles, one acute angle, and one obtuse angle.**

Activity 1

Quadrilaterals in the Classroom

Look around the room for quadrilaterals. Find examples of at least three different types of quadrilaterals illustrated in the beginning of this lesson. Draw each example you find, and next to each picture, name the object you drew and its shape. Then describe how you know that the object is the shape you named and describe the relationships of the sides of each quadrilateral.

Activity 2

Symmetry and Quadrilaterals

Materials needed:
- **Lesson Activity 25**
- mirror or reflective surface

If a figure can be divided into mirror images by a line of symmetry, then the figure has reflective symmetry. A mirror can help us decide if a figure has reflective symmetry. If we place a mirror upright along a line of symmetry, the half of the figure behind the mirror appears in the reflection of the other half. Use a mirror to discover which figures in **Lesson Activity 25** have reflective symmetry. If you find a figure with reflective symmetry, draw its line (or lines) of symmetry.

Lesson Practice

(**Classify**) Describe each quadrilateral as a trapezoid, parallelogram, rhombus, rectangle, or square. (More than one description may apply to each figure.)

a. 　　b. 　　c. 　　d.

e. Describe the angles in figures **a–d** and the relationships between the sides.

f. Draw two parallel line segments that are the same length. Then make a quadrilateral by drawing two more parallel line segments that connect the endpoints. Is your quadrilateral a parallelogram? Why or why not?

*1. **(Analyze)** What is the total number of days in the first three months of
(RF12) a leap year?

*2. Thirty-two desks were arranged as equally as possible in 6 rows.
(87)
    a. How many rows had exactly 5 desks?

    b. How many rows had 6 desks?

*3. **(Evaluate)** If $y = 3x + 6$, then what is y when $x = 4, 5,$ and 6?
(64,
Inv. 7)

*4. **(Analyze)** Carmen separated the 37 math books as equally as possible
(87) into 4 stacks.

    a. How many stacks had exactly 9 books?

    b. How many stacks had 10 books?

*5. **(Conclude)** Write *true* or *false* for parts **a–e.**
(90)
    a. All rectangles have four right angles.

    b. Some squares are rectangles.

    c. All trapezoids are rhombuses.

    d. All rectangles are parallelograms.

    e. Some parallelograms have no right angles.

6. a. What decimal number names the shaded part of the
(Inv. 4) large square at right?

    b. What decimal number names the part that is not
shaded?

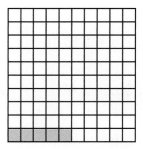

*7. **(Explain)** Near closing time, 31 children and adults are waiting in
(87) line to board a ride at an amusement park. Eight people board the
ride at one time. How many people will be on the last ride of the day?
Explain your answer.

8. Round 3874 to the nearest thousand.
(40)

9. ⟨Estimate⟩ Alicia opened a liter of milk and poured half of it into
(73) a pitcher. About how many milliliters of milk did she pour into the
pitcher?

10. The sun was up when Mark started working. It was dark when he
(13) stopped working later in the day. How much time had gone by?

Started   Stopped

*11. For five days Pilar recorded the high temperature. The temperatures
(Inv. 6) were 79°F, 82°F, 84°F, 81°F, and 74°F. What was the median temperature
for those five days?

12. ⟨Explain⟩ Leena drove 368 miles in 8 hours. If she drove the same
(60, 69) number of miles each hour, how far did she drive each hour? Explain
how you found your answer.

13.   496,325
(51) $+$ 3,680

14. $36.00
(52) $-$ $30.78

15. $12.45
(11, 28) $ 1.30
         $ 2.00
         $ 0.25
         $ 0.04
         $ 0.32
     $+$ $ 1.29

*16.   26
(88) $\times$ 24

*17.   25
(88) $\times$ 25

18. $8m = 16.40
(34, 83)

19. 60×300
(85)

20. 8.56×7
(59)

21. $7\overline{)845}$
(83)

22. $9\overline{)1000}$
(79)

23. $\dfrac{432}{6}$
(69)

24. ⟨Represent⟩ Draw and shade a circle that shows that $\frac{4}{4}$ equals 1.
(89)

25. The wall was 8 feet high and 12 feet wide. How many square feet of
(Inv. 3) wallpaper were needed to cover the wall?

26. **Analyze** Below are Tene's scores on the first seven games. Refer to
(Inv. 6) these scores to answer parts **a–c.**

$$85, 85, 100, 90, 80, 100, 85$$

    a. Rearrange the scores so that the scores are in order from lowest to
        highest.

    b. In your answer to part **a,** which score is the median score in
        the list?

    c. In the list of game scores, which score is the mode?

*27. **Estimate** What is a reasonable estimate of the number in each
(61) group when 912 objects are separated into 3 equal groups? Explain
why your estimate is reasonable.

*28. According to many health experts, a person should drink 64 ounces
(53) of water each day. If Shankeedra's glass holds 8 ounces of water, how
many glasses of water should she drink in one day?

*29. Arthur told his classmates that his age in years is a single-digit odd
(56) number greater than one. He also told his classmates that his age is
not a prime number. How old is Arthur?

*30. If $y = 3x - 1$, what is y when x is 2?
(64)

California Mathematics Content Standards

MG **2.0**, **2.1** Draw the points corresponding to linear relationships on graph paper (e.g., draw 10 points on the graph of the equation y = 3x and connect them by using a straight line).

MG **2.0**, **2.2** Understand that the length of a horizontal line segment equals the difference of the x-coordinates.

MG **2.0**, **2.3** Understand that the length of a vertical line segment equals the difference of the y-coordinates.

MR 2.0, 2.2 Apply strategies and results from simpler problems to more complex problems.

Focus on

Analyzing Relationships

In Investigation 7, we learned how to count segments or subtract coordinates to find the length of a side of a polygon that is drawn on a coordinate graph. In this investigation, we will subtract coordinates to find the length of a segment.

For segment *PS* below, we see that the coordinates of each *endpoint* are given. Because the coordinates are given, we can find the length of the segment without seeing the numbers on the *x*-axis and *y*-axis. Since segment *PS* is horizontal, the *y*-coordinates are always 6. So, we subtract the *x*-coordinates to find its length.

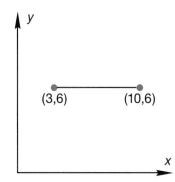

The *x*-coordinates of the endpoints are 10 and 3. Since $10 - 3 = 7$, the length of segment *PS* is 7 units.

1. (**Explain**) What is the length of segment *TR*? Explain your thinking.

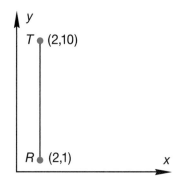

In Investigation 7 we also learned how to graph an equation such as $y = 2x + 1$. In the graph below, we can see that the line passes through the *y*-axis at 1. We can also see that the line $y = 2x + 2$ passes through the *y*-axis at 2.

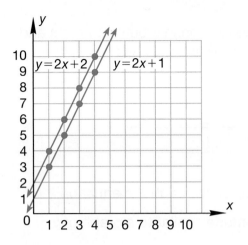

2. **Predict** Where will the line $y = 2x + 3$ cross the y-axis?

3. **Model** Write a set of ordered pairs for $y = 2x + 3$. Then use **Lesson Activity 26** to graph the line and check your prediction in problem **2.**

We have already studied how to make ordered pairs for a horizontal line such as $y = 5$. Another type of line we can make ordered pairs for is a vertical line. This line is the graph of $x = 5$.

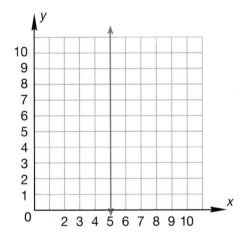

For every value of y, the value of x will always be 5. We can make a list of ordered pairs for this line.

$$x = (0 \cdot y) + 5$$

x	y
5	1
5	2
5	3

The table tells us that the points (5,1), (5, 2), and (5, 3) are points on the line $x = 5$.

4. **Connect** Write the coordinates of another point that is on this line.

5. **Discuss** Do the coordinates (6, 3) represent a point on this line? Explain why or why not.

6. **Model** Draw the graph of $x = 7$. Name the coordinates of three points on the line.

We have been writing equations to show the relationship of two quantities. Graphs can also be used to display relationships of two quantities.

7. (Generalize) This table shows the relationship between pints and cups. Complete the table. Then write a formula that can be used for changing pints to cups. In your formula, use c for cups and p for pints.

Pints	1	2	3	4	5
Cups	2	4			

8. (Represent) Use **Lesson Activity 26** to graph the equation.

9. (Discuss) How is the graph you made similar to a graph of $y = 2x$?

10. Name a relationship of two measures that can be graphed using an equation similar to $y = 2x$.

Investigate Further

a. Kanoni received a flat fee of $9 when she planted flowers, plus an additional $3 per hour for every hour she planted flowers. How much would she earn if it took her 5 hours to plant flowers?

Copy and extend the table to the right to find Kanoni's earnings.

Hours Worked	Money Earned
1	12
2	15
3	
4	
5	

Use the table to make a graph showing the relationship between hours worked and money made. Write the equation that the graph represents.

LESSON
91

• **Estimating Multiplication and Division Answers**

California Mathematics Content Standards

NS 1.0, **1.3** Round whole numbers through the millions to the nearest ten, hundred, thousand, ten thousand, or hundred thousand.

NS **3.0**, **3.4** Solve problems involving division of multidigit numbers by one-digit numbers.

MR 2.0, 2.1 Use estimation to verify the reasonableness of calculated results.

MR 2.0, 2.6 Make precise calculations and check the validity of the results from the context of the problem.

facts   Power Up I

mental math   Find half of each number in problems **a–d.**

  a. Number Sense: 24

  b. Number Sense: 50

  c. Number Sense: 46

  d. Number Sense: 120

  e. Money: The apples cost $3.67. Lindsay paid for them with a $5 bill. How much change should she receive?

  f. Estimation: About how many feet is 298 yards? (*Hint:* Round the number of yards to the nearest hundred yards before mentally calculating.)

  g. Calculation: $6 \times 7, -2, +30, +5, \div 3$

  h. Simplify: $4 - 2 \div 2$

problem solving   Choose an appropriate problem-solving strategy to solve this problem. There were two gallons of punch for the class party. The punch was served in 8-ounce cups. Two gallons of punch was enough to fill how many cups? (Remember that 16 ounces is a pint, two pints is a quart, two quarts is a half gallon, and two half gallons is a gallon.)

Estimation can help prevent mistakes. If we estimate the answer, we can tell whether our answer is reasonable.

Example 1

Luke multiplied 43 by 29 and got 203. Is Luke's answer reasonable?

We estimate the product of 43 and 29 by multiplying the rounded numbers 40 and 30.

$$40 \times 30 = 1200$$

Luke's answer of 203 and our estimate of 1200 are very different, so Luke's answer is **not reasonable.** He should check his work.

(Discuss) What is the exact product? Is the exact product close to 1200?

Example 2

An auditorium has 42 rows of seats. There are 53 seats in each row. The sales department sold 2000 tickets. Will all the people with a ticket have a seat or will they need to bring in extra chairs?

Since we do not need an exact answer, we can estimate the product by rounding each number to the nearest ten and then multiplying.

42 rounds to 40
53 rounds to 50

Now we can multiply. $40 \times 50 = 2000$

There are about 2000 seats in the auditorium. **All the people with tickets will have a seat.**

(Discuss) How can we decide if the estimate is reasonable?

Example 3

One company sent 6 trucks with a delivery of 1845 chairs for a new hotel. Estimate the number of chairs each truck will carry.

Thinking Skills

Estimate

How would you estimate the quotient of 184 ÷ 6?

We choose a number close to 1845 that is easily divided by 6. We know that 18 is a multiple of 6, so 1800 is a compatible dividend. We can calculate mentally: "18 hundred divided by 6 is 3 hundred."

$$1800 \div 6 = 300$$

Each truck will carry about **300 chairs.**

Lesson Practice   Estimate each product or quotient. Then find the exact answer.

a. 58 × 23   **b.** 49 × 51   **c.** 61 × 38   **d.** 1845 ÷ 9

e. **Estimate** A ferry can carry 843 people. About how many people can the ferry carry on a round trip?

Written Practice   *Distributed and Integrated*

*1. **Interpret** Use the information in this circle graph to answer parts **a–d.**
(Inv. 5)

Activities of 100 Children at the Park

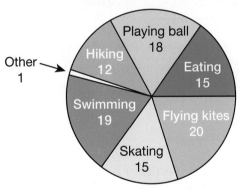

a. Altogether, how many children were at the park?

b. How many children were not swimming?

c. How many children were either hiking or skating?

d. How many more children were flying kites than were swimming?

*2. **Justify** Name a prime number that is a factor of 115. Explain how you
(56) know it is prime.

*3. **(Analyze)** Simplify each expression. Remember to use the order of
(63) operations.

     a. $4 + (15 - 5) \div 2$ **b.** $(36 \div 3) - (2 \times 4)$

4. Write each mixed number as a decimal:
(Inv. 4)

     a. $3\frac{5}{10}$ **b.** $14\frac{21}{100}$ **c.** $9\frac{4}{100}$

*5. Estimate the product of 39 and 406. Then find the exact product.
(91)

*6. If $y = 4x - 2$, what is y when x is 4?
(64)

*7. Write these fractions in order from least to greatest:
(57)

                            $\frac{3}{4}$ $\frac{1}{2}$ $\frac{5}{8}$

8. Compare: 2 thousand $\bigcirc$ 24 hundreds
(27)

Refer to the rectangle below to answer problems **9** and **10**.

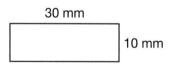

30 mm

10 mm

9. What is the perimeter of the rectangle
(20, 42)
     a. in millimeters?

     b. in centimeters?

10. What is the area of the rectangle
(42, 66)
     a. in square millimeters?

     b. in square centimeters?

11. Santos figured the trip would take seven and a half hours. He left
(13) at 7 a.m. At what time does he think he will arrive?

*12. **(Conclude)** Write *true* or *false* for parts **a–e.**
(70, 81,
  90)   **a.** All squares are rectangles.

     b. All trapezoids are quadrilaterals.

     c. All equilateral triangles are congruent.

     d. All triangles have at least one right angle.

     e. All squares are similar.

13. 25×40
(71)

14. $98¢ \times 7$
(28, 38)

15. $\sqrt{36} \times \sqrt{4}$
(Inv. 3)

16. $\dfrac{3^3}{3}$
(65)

17. $\begin{array}{r} 36 \\ \times\ 34 \\ \hline \end{array}$
(88)

18. $\begin{array}{r} 35 \\ \times\ 35 \\ \hline \end{array}$
(88)

19. $\begin{array}{r} 4 \\ 2 \\ 1 \\ 3 \\ 4 \\ 7 \\ 2 \\ 2 \\ 3 \\ 4 \\ +\ x \\ \hline 42 \end{array}$
(4)

20. $8m = \$70.00$
(34, 79)

21. $6\overline{)1234}$
(83)

22. $800 \div 7$
(79)

23. $487 \div 3$
(79)

24. $\$2.74 + \$0.27 + \$6 + 49¢$
(28)

25. $9.48 - (3.7 + 2.36)$
(9, 45)

* 26. (**Represent**) Draw and shade circles to show that $2\frac{1}{3}$ equals $\frac{7}{3}$.
(89)

* 27. (**Analyze**) Listed below are the number of points Amon scored in his
(Inv. 6) last nine basketball games, which range from 6 to 10. Refer to these scores to answer parts **a–c.**

$$8, 7, 7, 8, 6, 10, 9, 10, 7$$

   a. What is the mode of the scores?

   b. What is the median of the scores?

   c. Are there any outliers?

28. Each school day, Brent's second class begins at 9:00 a.m. What
(17) kind of angle is formed by the minute hand and the hour hand of a clock at that time?

* 29. (**Explain**) Thirty-one students are entering a classroom. The desks
(87) in the classroom are arranged in rows with 7 desks in each row. If the students fill the first row of desks, then fill the second row of desks, and so on, how many full rows of students will there be? How many students will sit in a row that is not full? Explain your answer.

*30. (Inv. 9) **Conclude** What is the length of this segment? Explain your reasoning.

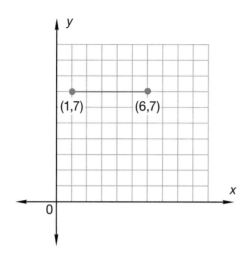

Real-World Connection

Mrs. Collins has to make some salads for 4 friends and herself. She has two salad recipes to choose from—one for a *garden salad* and one for a *spinach salad*. She has the following ingredients to work with: 24 tomatoes, 30 mushroom slices, 16 ounces of carrots, 36 ounces of lettuce, and 50 ounces of spinach.

Garden Salad for 1 person:	**Spinach Salad for 1 person:**
4 mushroom slices	6 mushroom slices
1 ounce of carrots	3 ounces of carrots
4 tomatoes	5 tomatoes
7 ounces of lettuce	9 ounces of spinach

a. Does Mrs. Collins have enough ingredients to make 5 *garden salads*?

b. Which ingredient does Mrs. Collins not have enough of to make 5 *spinach salads*?

c. Does Mrs. Collins have enough ingredients to make 3 *spinach salads* and 2 *garden salads*?

California Mathematics Content Standards

NS 1.0, 1.6 Write tenths and hundredths in decimal and fraction notations and know the fraction and decimal equivalents for halves and fourths (e.g., $\frac{1}{2} = 0.5$ or $.50$; $\frac{7}{4} = 1\frac{3}{4} = 1.75$).

NS 1.0, 1.7 Write the fraction represented by a drawing of parts of a figure; represent a given fraction by using drawings; and relate a fraction to a simple decimal on a number line.

NS 1.0, **1.9** Identify on a number line the relative position of positive fractions, positive mixed numbers, and positive decimals to two decimal places.

• Comparing and Ordering Fractions and Decimals

facts   Power Up I

mental math   Find half of a product in problems **a–c.**

   a. Number Sense: half of 10×12

   b. Number Sense: half of 10×24

   c. Number Sense: half of 10×480

   d. Money: The art supplies cost $17.50. Adam paid with a $20 bill. How much change should he receive?

   e. Estimation: About what fraction of the circle is shaded? About what fraction of the circle is not shaded?

   f. Calculation: $\frac{1}{4}$ of 40, $\times\, 2$, $+\, 4$, $\div\, 3$

   g. Simplify: $5 + 3 \times 3$

problem solving   Choose an appropriate problem-solving strategy to solve this problem. Below we show the first five terms of a sequence. The terms of the sequence increase from left to right. Estimate how many terms will be in the sequence when it reaches a number that is 500 or greater. Then check your estimate by continuing the sequence until you reach a number that is 500 or greater.

$$1, 2, 4, 8, 16, \ldots$$

Since Lesson 57, we have been comparing and ordering fractions with different denominators using pictures and diagrams. We can also use number lines to compare and order fractions with different denominators.

We can use this number line to write the fractions $\frac{1}{2}$, $\frac{2}{5}$, and $\frac{7}{10}$ in order from least to greatest. We can use $\frac{5}{10}$ as a benchmark for comparing all three fractions.

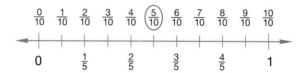

We know that $\frac{5}{10}$ is one half of a whole region.

$$\frac{5}{10} = \frac{1}{2}$$

On the number line we can see that $\frac{2}{5}$ is equivalent to $\frac{4}{10}$, so $\frac{2}{5}$ is less than $\frac{1}{2}$.

We can also see that $\frac{7}{10}$ is greater than $\frac{1}{2}$.

The fractions written in order from least to greatest are:

$$\frac{2}{5}, \frac{1}{2}, \frac{7}{10}$$

Example 1

Use this number line to write these fractions in order from greatest to least:

$$\frac{3}{8}, \frac{1}{2}, \frac{3}{4}$$

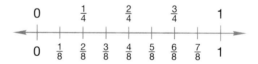

We can see on this number line that $\frac{2}{4}$ and $\frac{4}{8}$ are halfway between 0 and 1. This means that both $\frac{2}{4}$ and $\frac{4}{8}$ are equal to $\frac{1}{2}$. We can use this fact to compare all three numbers.

$$\frac{3}{8} < \frac{1}{2}, \quad \frac{3}{4} > \frac{1}{2}$$

The fractions written in order from greatest to least are:

$$\frac{3}{4}, \frac{1}{2}, \frac{3}{8}$$

Verify What number did we use as a benchmark for comparing?

Example 2

Use this number line to write these fractions and decimals in order from least to greatest.

$$\frac{1}{3}, 0.5, 0.25, \frac{4}{6}$$

We can see on this number line that $\frac{3}{6}$ is halfway between 0 and 1. This means that $\frac{3}{6}$ is equal to $\frac{1}{2}$.

Since $0.5 = \frac{1}{2}$ and $\frac{1}{3}$ is less than $\frac{1}{2}$, then $\frac{1}{3}$ is less than 0.5.

Since $\frac{4}{6}$ is greater than $\frac{3}{6}$, then $\frac{4}{6}$ is greater than 0.5.

Now, we write $\frac{4}{6} > 0.5 > \frac{1}{3}$.

If we change 0.25 to the fraction $\frac{1}{4}$, we can compare $\frac{1}{3}$ and $\frac{1}{4}$ and we know that $\frac{1}{3} > \frac{1}{4}$.

These numbers written in order from least to greatest are:

$$0.25, \frac{1}{3}, 0.5, \frac{4}{6}$$

Discuss Why did we use $\frac{1}{2}$ as a benchmark for comparing?

Example 3

Write the letter that represents each on this number line:

a. $1\frac{3}{4}$ b. 1.50   c. $\frac{1}{2}$ d. 2.75

On this number line, there are four segments between 0 and 1, 1 and 2, and 2 and 3. This means that this number line is divided into fourths.

a. Since each unit segment equals $\frac{1}{4}$, we count three segments after 1 and see that the letter **G** represents $1\frac{3}{4}$.

b. The letter *F* is halfway between 1 and 2, so **F** represents 1.50.

c. The letter *B* is halfway between 0 and 1, so *B* represents $\frac{1}{2}$.

d. We count three segments after 2, so the letter *K* represents 2.75.

Connect Use this number line to compare $\frac{7}{4}$ and 2.30.

Lesson Practice

Analyze Use this number line to answer problems **a–d**.

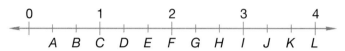

a. Which letter represents $1\frac{2}{3}$?

b. Which letter represents $\frac{1}{3}$?

c. Which letter represents the number closest to 2.92?

d. Which letter represents $\frac{10}{3}$?

Verify Write *true* or *false* for each problem. Use the number lines in this lesson to help you.

e. $\frac{5}{3} > 1\frac{1}{3}$ $\qquad$ **f.** $\frac{6}{4} < 2.5$

g. $\frac{3}{5} > 0.75$ $\qquad$ **h.** $1\frac{2}{8} < \frac{11}{8}$

Written Practice

Distributed and Integrated

＊1. **Represent** Draw and shade circles to show that $\frac{10}{4} = 2\frac{1}{2}$.
(89)

＊2. **Analyze** Write these numbers in order from greatest to least. Use the
(92) number line on page 611 if needed.

$$\frac{3}{5}, \, 1.01, \, \frac{6}{5}$$

3. Compare. Write >, <, or =.
(27)
$\quad$ **a.** 206,353 $\bigcirc$ 209,124 $\qquad$ **b.** 518,060 $\bigcirc$ 518,006

4. Write these numbers in order from greatest to least:
(27)
$\qquad$ 89,611 $\qquad$ 120,044 $\qquad$ 102,757 $\qquad$ 96,720

5. **Represent** Write each mixed number as a decimal:
(Inv. 4, 43)
$\quad$ **a.** $5\frac{31}{100}$ $\qquad$ **b.** $16\frac{7}{10}$ $\qquad$ **c.** $5\frac{7}{100}$

*6. **Explain** 33 people are waiting for a boat ride. Each boat holds 6 people.
(87) If five boats arrive together, how many people will have to wait for the sixth boat? How did you find your answer?

*7. **Analyze** Jim spun all the way around in the air and dunked the
(78) basketball. How many degrees did Jim turn?

8. **Represent** Use words to write 7.68.
(Inv. 4)

9. **Represent** Use words to write 76.8.
(Inv. 4)

*10. ✎ **Explain** Armando estimated that the exact product of 78 and 91
(91) was close to 720. Did Armando make a reasonable estimate? Explain why or why not.

11. **Connect** Name the number of shaded squares below:
(Inv. 4)
    a. as a mixed number.

    b. as a decimal.

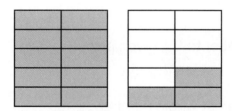

*12. **Represent** Draw a quadrilateral with exactly two right angles.
(90)

13. Makayla's school day ends 5 hours 20 minutes after the
(13) time shown on the clock. What time does Makayla's school day end?

14. Mr. Romano could bake 27 pizzas in 3 hours.
(58, 60)
    a. How many pizzas could he bake in 1 hour?

    b. How many pizzas could he bake in 5 hours?
      (*Hint:* Multiply the answer to part **a** by 5.)

15. $3.65 + 4.2 + 0.62$
(45)

16. $\$13.70 - \$6.85$
(28, 52)

17. 26×100
(85)

18. $9 \times 87¢$
(28, 38)

19. 14×16
(88)

20. 15^2
(65, 88)

21. $\dfrac{456}{6}$
(69)

22. $\begin{array}{r} 47 \\ \times\ \ 60 \\ \hline \end{array}$
(71)

23. $6x = 4248$
(34, 83)

24. $1\overline{)163}$
(79)

25. $5\overline{)\$49.00}$
(79, 83)

*26. This table represents the equation $y = 2x + 3$ and shows the values of y when x is 2 and when x is 3. What is y when x is 4?
(64, Inv. 7)

$y = 2x + 3$	
x	**y**
2	7
3	9
4	?

27. How many one-foot-square tiles are needed to cover the floor of a room that is 15 feet long and 10 feet wide?
(Inv. 3, 84)

*28. Find the median and mode of this set of numbers:
(Inv. 6)

$$1, 1, 2, 3, 5, 8, 13$$

*29. (Estimate) Round each number to the given place:
(22, 46, 62)

   a. Round 65.25 to the nearest ten.

   b. Round 65.25 to the nearest whole number.

   c. Round 65.25 to the nearest tenth.

*30. (Explain) If $p + 50 = r + 50$, is the equation below true? Why or why not?
(63)

$$p \div 2 = r \div 2$$

California Mathematics Content Standards

NS **3.0, 3.4** Solve problems involving division of multidigit numbers by one-digit numbers.

AF 1.0, **1.5** Understand that an equation such as $y = 3x + 5$ is a prescription for determining a second number when a first number is given.

MR 2.0, 2.5 Indicate the relative advantages of exact and approximate solutions to problems and give answers to a specified degree of accuracy.

• Two-Step Problems

Power Up

facts   Power Up I

mental math   Find half of a product in problems **a–c.**

    a. Number Sense: half of 10×18

    b. Number Sense: half of 10×44

    c. Number Sense: half of 10×260

    d. Time: How many minutes are in $1\frac{1}{2}$ hours?

    e. Measurement: How many quarts is 3 gallons?

    f. Estimation: About how many feet is 1989 yards?

    g. Calculation: 3^2, $+ 1$, $\times 5$, $- 1$, $\sqrt{}$

    h. Simplify: $6 - 3 \div 3$

problem solving   In this Venn diagram, the circle on the left represents animals with the ability to fly, and the circle on the right represents birds. The R in the overlapping portion of the circles represents robins, which are birds that can fly. The O represents ostriches, which are birds that cannot fly. The B represents bats, which can fly but are not birds. The W represents whales, which are not birds and cannot fly. Copy the Venn diagram on your paper, and place an abbreviation for a penguin, eagle, goldfish and cat.

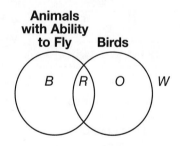

We have practiced two-step word problems that involve finding total costs (including tax) and change back. Starting with this lesson, we will practice other kinds of two-step problems. Writing down the given information and using problem-solving strategies is often helpful in solving these problems.

Example 1

Reading Math

When we translate a problem, we identify the goal and list the steps.

Goal: Find Jim's age.

Step 1: Find Ali's age.

Step 2: Find Jim's age.

Then we use the steps to make a plan.

Jim is 5 years older than Ali. Ali is 2 years younger than Blanca. Blanca is 9 years old. How old is Jim?

We will use two steps to solve the problem. First we will use Blanca's age to find Ali's age. Then we will use Ali's age to calculate Jim's age. We write down the given information.

> Blanca is 9 years old.
>
> Ali is 2 years younger than Blanca.
>
> Jim is 5 years older than Ali.

We know that Blanca is 9 years old. Ali is 2 years younger than Blanca, so Ali is $9 - 2$, or 7 years old. Jim is 5 years older than Ali, so Jim is $7 + 5$, or **12 years old.**

Example 2

Thinking Skills

Verify

What are the two steps needed to find the cost of each pound?

Ja'Von paid for 5 pounds of apples with a $10 bill. His change was $6. What was the cost of each pound of apples?

We begin by finding how much all 5 pounds of apples cost. If Ja'Von paid for the apples with a $10 bill and received $6 in change, then all 5 pounds must have cost $4.

$$\begin{array}{r} \$10 \\ -\ \$\ 6 \\ \hline \$\ 4 \end{array} \quad \begin{array}{l} \text{amount paid} \\ \text{change} \\ \text{cost of 5 pounds of apples} \end{array}$$

To find the cost of each pound of apples, we divide $4 by 5.

$$\begin{array}{r} \$0.80 \\ 5\overline{)\$4.00} \\ \underline{4\ 0} \\ 00 \\ \underline{0} \\ 0 \end{array}$$

Each pound of apples cost **$0.80.**

Example 3

Maribella feeds her pet rabbit 2 ounces of lettuce each day. In how many days does her rabbit eat a pound of lettuce? How many pounds of lettuce does the rabbit eat in 4 months?

A pound is 16 ounces. At 2 ounces per day, the rabbit eats a pound of lettuce every **8 days.**

$$16 \div 2 = 8$$

A month is about 30 days, so 4 months is 4×30 days, which is 120 days. We divide 120 days into groups of 8 days to find the number of pounds of lettuce the rabbit eats.

$$120 \div 8 = 15$$

In 4 months, the rabbit eats about **15 pounds** of lettuce.

Example 4

Point *B* represents which number on this number line?

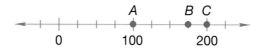

Sometimes two-step problems are not word problems. We can solve problems like this with two or three steps of arithmetic.

We see that the distance from point *A* to point *C* is 100.

Step 1: $200 - 100 = 100$

The distance is divided into 4 segments. By dividing 100 by 4, we find that each segment is 25.

Step 2: $100 \div 4 = 25$

Step 3: If we count by 25's from 100, point *A* to point *B*, we find that point *B* represents **175.** Since point *B* is one segment from point *C,* we can check the answer by counting back 25 from 200. The result is 175, which is our original answer.

(**Discuss**) How could we use the *guess and check* strategy to solve this problem?

Example 5

If $y = 2x + 1$, then what is y when $x = 3$?

The equation $y = 2x + 1$ shows us how to find the number that y equals when we know what x equals.

The equation means, "To find y, multiply x by 2 and then add 1."

In this equation x is 3, so we multiply 2 times 3 and then add 1.

$$y = (2 \times 3) + 1$$
$$y = \quad 6 \quad + 1$$
$$y = 7$$

When x is 3, y is **7**.

Represent What is y when x is 5?

We can write these values in a table to find the answer.

$2x + 1 = y$

x	y
3	7
4	9
5	11

When x is 5, y is **11**.

Predict What is y when $x = 10$? Explain how you know.

Lesson Practice

a. Kim paid for 4 pounds of peaches with a $5 bill. She got $3 back. What was the cost of each pound of peaches? (*Hint:* First find the cost of 4 pounds of peaches.)

b. The perimeter of this square is 12 inches. What is the area of the square? (*Hint:* First find the length of each side.)

c. Orlando is 10 years younger than Gihan and Gihan is 2 years older than Shaniqua. If Orlando is 13 years old, how old is Shaniqua? (*Hint:* First find how old Gihan is.)

d. Point N represents what number on this number line?

e. If $y = 3x + 2$, what is y when x is 4?

f. Mr. Simmons is 5 ft 10 in. tall. How many inches is 5 ft 10 in.?

*1. Fifty-three family photographs are being arranged in a photo album.
(87) The album has 12 pages altogether, and 6 photographs can be placed on each page.

    a. How many full pages of photographs will be in the album?

    b. How many photographs will be on the page that is not full?

    c. How many pages in the album will be empty?

2. (**Estimate**) Abraham Lincoln was born in 1809. He gave the
(61) Gettysburg Address in 1863. About how old was he when he gave the Gettysburg Address?

*3. (**Analyze**) The parking lot charges $1.25 to park a car for the first hour.
(93) It charges 75¢ for each additional hour. How much does it cost to park a car in the lot for 3 hours?

*4. (**Represent**) Two thirds of the team's 45 points were scored in the
(74) second half. How many points did the team score in the second half? Draw a picture to illustrate the problem.

*5. Something is wrong with the sign to the right. Show two
(28) different ways to correct the error.

6. (**Analyze**) What is the value of 3 $10 bills, 4 $1 bills, 5 dimes, and
(28) 2 pennies?

7. (**Represent**) Use words to write 6412.5, and then round it to the
(40, nearest thousand.
Inv. 4)

8. **Estimate** Last year, 5139 people attended an outdoor jazz
(61) festival. This year, 6902 people attended the festival. Estimate the
total attendance during those years and explain why your estimate
is reasonable.

9. a. Cooper opened a 1-gallon bottle of milk and poured out 1 quart.
(73, 77) How many quarts of milk were left in the bottle?

b. What fraction of the milk was left in the bottle?

* **10.** **Multiple Choice** Which of the following figures has exactly 4 lines of
(82) symmetry? Copy the figure and draw the lines of symmetry.

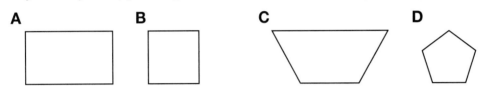

A   B   C   D

* **11.** Estimate the product of 39 and 41. Then find the exact product.
(91)

* **12.** **Estimate** Felicia slowly gave the doorknob a quarter turn
(78) counterclockwise. About how many degrees did she turn the
doorknob?

* **13.** **Explain** Write these numbers in order from greatest to least. Explain
(92) your thinking.

$$\frac{6}{5}, \ 1.55, \ 1\frac{2}{4}$$

14. $68.57
(28, 51) + $36.49

15. $100.00
(28, 52) − $ 5.43

16.   15
(11)   24
           36
           75
           21
            8
           36
       + 420

17.   12
(86) × 12

18. $5.08
(59) ×   7

19. 50^2
(65, 85)

20. $\sqrt{144}$
(Inv. 3)

21. $12.08 - (9.61 - 2.4)$
(9, 45)

22. 49×51
(88)

23. 33×25
(88)

24. $\frac{848}{8}$
(83)

25. $9w = 6300$
(34, 83)

*26. (**Represent**) Draw and shade circles to show that $2\frac{2}{3}$ equals $\frac{8}{3}$.
(89)

*27. (**Represent**) Draw a rectangle that is three inches long and one inch
(18, 20, 66) wide. Then find the perimeter and the area.

*28. This table represents the equation $y = 3x + 1$ and shows
(64, Inv. 7) the values of y when x is 3 and when x is 4.

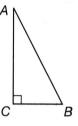

y = 3x + 1	
x	**y**
3	10
4	13
5	?

    a. What is y when x is 5?

    b. Use **Lesson Activity 26** to graph the equation.

*29. (**Classify**) Refer to this triangle for parts **a** and **b**.
(17)

    a. Describe the angles as acute, right, or obtuse.

    b. Which sides are perpendicular?

*30. (**Classify**) Write as many different names for this figure
(90) as you can.

LESSON

94

✎ *California Mathematics Content Standards*

NS 1.0, 1.5 Explain different interpretations of fractions, for example, parts of a whole, parts of a set, and division of whole numbers by whole numbers; explain equivalents of fractions (see Standard 4.0).

NS 1.0, 1.7 Write the fraction represented by a drawing of parts of a figure; represent a given fraction by using drawings; and relate a fraction to a simple decimal on a number line.

MR 1.0, 1.2 Determine when and how to break a problem into simpler parts.

• Two-Step Problems About a Fraction of a Group

facts	Power Up H
mental math	Five is half of 10. To multiply by 5, we can multiply by half of 10. For example, 5×12 equals half of 10×12. Find each product by multiplying by "half of 10" in problems **a–d.**

    a. Number Sense: 5×16

    b. Number Sense: 5×24

    c. Number Sense: 5×28

    d. Number Sense: 5×64

    e. Measurement: A *stone* is a British unit of weight equal to 14 pounds. Two stone is 28 pounds, 3 stone is 42 pounds, and so on. How many pounds is 10 stone?

    f. Estimation: Lydia walked 1 km in 608 seconds. About how many minutes did it take her to walk 1 km?

    g. Calculation: $\frac{1}{10}$ of 40, $\times$ 10, $+$ 5, $\div$ 5

    h. Simplify: $8 + 2 \times 2$

problem solving	Choose an appropriate problem-solving strategy to solve this problem. Find the next five numbers in this sequence. Then describe the sequence in words.

$$\ldots, 64, 32, 16, 8, \underline{\quad}, \underline{\quad}, \underline{\quad}, \underline{\quad}, \underline{\quad}, \ldots$$

New Concept

The word problems in this lesson are two-step problems involving fractions of a group. First we divide to find the number in one part. Then we multiply to find the number in more than one part.

Example 1

Thinking Skills

Verify

What are the two steps needed to find the number of campers who wore green jackets?

There were 30 campers in the state park. Two thirds of them wore green jackets. How many campers wore green jackets?

The word *thirds* tells us there were 3 equal groups. First we find the number of campers in each group. Since there were 30 campers in all, we divide 30 by 3.

$$\begin{array}{r} 10 \\ 3\overline{)30} \end{array}$$

There were 10 campers in each group. We draw this diagram:

30 campers

$\frac{2}{3}$ wore green jackets. $\left\{ \begin{array}{|c|} \hline 10 \text{ campers} \\ \hline 10 \text{ campers} \\ \hline \end{array} \right.$

$\frac{1}{3}$ did not wear green jackets. $\left\{ \begin{array}{|c|} \hline 10 \text{ campers} \\ \hline \end{array} \right.$

Two thirds wore green jackets. In two groups there were 2×10 campers or **20 campers** who wore green jackets. We also see that one group did not wear green jackets, so 10 campers did not wear green jackets.

Example 2

The force of gravity on Mars is about $\frac{2}{5}$ the force of gravity on Earth. A rock brought back to Earth from Mars weighs 50 pounds. How much did the rock weigh on Mars?

The mass of the rock is the same on Earth as it was on Mars because it is the same amount of rock. However, Earth is more massive than Mars, so the force of gravity is greater on Earth. The rock on Mars weighed only $\frac{2}{5}$ of its weight on Earth. To find $\frac{2}{5}$ of 50 pounds, we first find $\frac{1}{5}$ of 50 pounds by dividing 50 pounds by 5.

$$50 \text{ pounds} \div 5 = 10 \text{ pounds}$$

50 pounds

weight of the rock on Mars $\left\{ \begin{array}{|c|} \hline 10 \text{ pounds} \\ \hline 10 \text{ pounds} \\ \hline 10 \text{ pounds} \\ \hline 10 \text{ pounds} \\ \hline 10 \text{ pounds} \\ \hline \end{array} \right.$

Each fifth is 10 pounds, so $\frac{2}{5}$ is 20 pounds. We find that the rock that weighs 50 pounds on Earth weighed only **20 pounds** on Mars.

Discuss Why did the weight of the rock change when it was brought to Earth? How does the mass of the rock on Mars compare to the mass of the rock on Earth?

Lesson Practice

Represent Diagram problems **a** and **b**. Then answer the question.

a. Three fourths of the 24 checkers were still on the board. How many checkers were still on the board?

b. Two fifths of 30 students studied more than one hour for a test. How many students studied for more than one hour?

c. The force of gravity on Mercury is about $\frac{1}{3}$ the force of gravity on Earth. How much would a tire weigh on Mercury, if it weighs 39 lb on Earth? Would the mass be the same? Why or why not?

Written Practice — Distributed and Integrated

1. One hundred fifty feet equals how many yards?
(42, 75)

2. Tammy gave the clerk $6 to pay for a book. She received 64¢ in change. Tax was 38¢. What was the price of the book?
(93)

* **3.** DaJuan is 2 years older than Rebecca. Rebecca is twice as old as Dillon. DaJuan is 12 years old. How old is Dillon? (*Hint:* First find Rebecca's age.)
(93)

4. Write each decimal as a mixed number:
(Inv. 4, 43)
  a. 3.29   **b.** 32.9   **c.** 3.09

* **5.** **Represent** **a.** Three fourths of the 84 contestants guessed incorrectly. How many contestants guessed incorrectly? Draw a picture to illustrate the problem.
(44, 94)

  b. What decimal number represents the contestants who guessed incorrectly?

*6. **(Analyze)** In February, Fairbanks, Alaska, has an average high
(21)   temperature of 7°F and an average low temperature of −14°F. How
       many degrees difference are these two temperatures?

7. a. What is the diameter of this circle?
(18, 42)

   b. What is the radius of this circle?

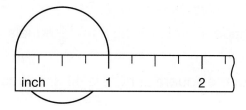

8. **(Represent)** Use words to write 8.75.
(Inv. 4)

*9. **(Estimate)** Three students each made a different estimate of the
(91)   quotient 2589 ÷ 9. Paulo's estimate was 30, Carter's estimate was
       300, and Allison's estimate was 3000. Which student made the best
       estimate? Explain your answer.

*10. The first five odd counting numbers are 1, 3, 5, 7, and 9. Find the
(Inv. 6)   median of these five numbers.

*11. **(Explain)** Write these numbers in order from least to greatest. Explain
(92)   your thinking.

$$\frac{7}{4}, \ 1.5, \ \frac{14}{10}$$

*12. **Multiple Choice** Use the polygons below to answer parts **a–c.**
(90)

A   B   C   D

   a. Which of these polygons is a parallelogram?

   b. Which polygon(s) appear to have at least one obtuse angle?

   c. Which polygon does not appear to have any perpendicular sides?

13. $16.25 − ($6 − 50¢)
(9, 28,
52)

14. 5 × 7 × 9   **15.** $7.83 × 6   **16.** 54 × 1000
(38)   (59)   (84)

17. 45
(88) × 45

18. 32
(71) × 40

19. 46
(88) × 44

20. 6)3625
(83)

21. 5)3000
(83)

22. $7n = 987$
(34, 79)

23. $\dfrac{10^3}{\sqrt{25}}$
(Inv. 3, 65, 83)

24. $\$13.76 \div 8$
(79)

25. $\dfrac{232}{4}$
(68)

* **26.** (Represent) Draw and shade a circle to show that $\frac{8}{8}$ equals 1.
(89)

* **27.** (Analyze) The perimeter of the square at right is 40 cm. What is the area of this square?
(31, 66)

28. (Represent) Draw a triangle that is similar to this isosceles triangle. Then draw its line of symmetry.
(70, 82)

* **29.** **a.** Compare: $0.25 \bigcirc 0.250$
(28, 44)

b. Compare: $\$0.25 \bigcirc 25¢$

* **30.** (Analyze) Madison had $100 in the bank. She wrote a check for $145. What number represents the balance of her account?
(76)

California Mathematics Content Standards

AF 1.0, **1.3**   Use parentheses to indicate which operation to perform first when writing expressions containing more than two terms and different operations.

SDAP 1.0, 1.1   Formulate survey questions; systematically collect and represent data on a number line; and coordinate graphs, tables, and charts.

SDAP 1.0, 1.2   Identify the mode(s) for sets of categorical data and the mode(s), median, and any apparent outliers for numerical data sets.

• Describing Data

facts   Power Up H

mental math

Find each product by multiplying by "half of 10" in problems **a–c.**

  a. Number Sense: 5×46

  b. Number Sense: 5×62

  c. Number Sense: 5×240

  d. Money: The price of the blouse is $24.87. Sales tax is $1.95. What is the total cost?

  e. Measurement: The large glass of water weighed half a kilogram. How many grams is half a kilogram?

  f. Estimation: The package of 10 pencils costs $1.98. Round that price to the nearest dollar and then divide by 10 to estimate the cost per pencil.

  g. Calculation: $\sqrt{4}$, $\times\, 7$, $+\, 1$, $+\, 10$, $\sqrt{}$, $-\, 4$

  h. Simplify: $10 - 6 \div 2$

problem solving

Choose an appropriate problem-solving strategy to solve this problem. On February 4, Edgar remembered that his two library books were due on January 28. The fine for late books is 15¢ per book per day. If he returns the books on February 4, what will be the total fine?

New Concept

The median of a set of numbers is the middle number when the numbers are arranged in order of size. When there is an even set of numbers, the median is the sum of the two middle numbers divided by two.

Find the median of Ian's seven game scores.

80, 85, 85, 10, 90, 90, 85

The median score is the middle score. To find the median score, we arrange the scores in order. We begin with the lowest score.

10, 80, 85, 85, 85, 90, 90

3 scores   ↑   3 scores

middle

We see that the median score is **85.**

Discuss Explain how to find the median of the following set of numbers.

5, 4, 3, 8, 7, 7

Notice that the low score of 10 does not affect the median. A score that is far from the other scores is called an outlier. Outliers generally have little or no effect on the median. Below we have placed these scores on a line plot. We see that most of the scores are close together.

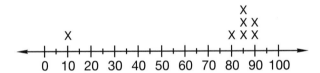

The outlier is far away from the other scores.

The **range** of a set of numbers is the difference between the largest and the smallest numbers in a list. To calculate the range of a list, we subtract the smallest number from the largest number.

Find the range of Ian's seven game scores.

80, 85, 85, 10, 90, 90, 85

The scores vary from a low of 10 to a high of 90. The range is the difference of the high and low scores. We subtract 10 from 90 and find that the range is **80.**

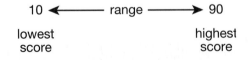

The mode of a set of numbers is the number that occurs most often.

Example 3

Find the mode of Ian's seven game scores.

80, 85, 85, 10, 90, 90, 85

We see that the score of 85 appears three times. No other score appears more than twice. So the mode is **85.**

Discuss What is the mode of the following set of numbers?

5, 5, 3, 6, 8, 7, 7

Collecting Data

Materials needed:
- paper and pencil

Use a sheet of paper for recording data from a survey. Then take a survey of ten friends. Use this question or one of your own:

How many bicycles do you have in your home?

a. Record the data in a table.

b. Use the data to make a line plot.

c. Write the median and mode.

Lesson Practice

a. **Analyze** Find the median, mode, and range of Raquel's game points shown below. Is there an outlier in this set of points?

50, 80, 90, 85, 90, 95, 90, 100

b. Find the median, mode, and range of this set of numbers:

31, 28, 31, 30, 25

c. **Explain** Find the median of these temperatures. Explain how you found your answer.

75°, 80°, 80°, 90°, 95°, 100°

d. **Interpret** Every X on this line plot stands for the age of a child attending a party. How many children attended the party? What is the mode of the ages?

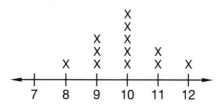

e. (Verify) Is there an outlier in the data in problem **d**? Why or why not?

Written Practice   *Distributed and Integrated*

*1. (Interpret) Use the information in this table to answer parts **a–c.**
(Inv. 5)

Average Yearly Rainfall

City	Rainfall (in inches)
Boston	43
Chicago	36
Denver	16
Houston	48
San Francisco	20

a. Which cities listed in the table average less than 2 feet of rain per year?

b. One year Houston received 62 inches of rain. This was how much more than its yearly average?

c. Copy and complete this bar graph to show the information in the rainfall table:

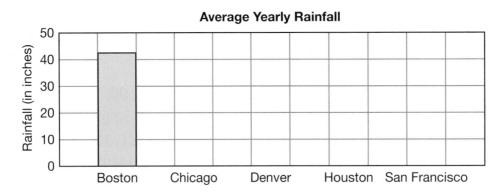

*2. (Represent) Five sixths of the 288 marchers were out of step. How
(94) many marchers were out of step? Draw a picture to illustrate the problem.

*3. (Represent) Something is wrong with this sign. Draw
(28) two different signs that show how to correct the error.

4. What is the radius of this circle in millimeters?
_(18, 42)

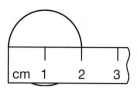

*5. (**Verify**) Write *true* or *false* for each parts **a–c.**
₍₉₂₎

    a. $\dfrac{7}{6} > \dfrac{7}{4}$

    b. $\dfrac{7}{10} > 0.5$

    c. $1\dfrac{4}{6} < 1\dfrac{5}{10}$

*6. Estimate the product of 88 and 22. Then find the actual product.
₍₉₁₎

7. Apples were priced at 53¢ per pound. What was the cost of 5 pounds
₍₃₉₎ of apples?

8. (**Represent**) Write the number 3708 in expanded form. Then use words
₍₂₇₎ to write the number.

9. The top of a doorway is about two meters from the floor. Two meters is
₍₄₂₎ how many centimeters?

*10. Four pounds of pears cost $1.20. What did 1 pound of pears cost?
₍₉₃₎ What did 6 pounds of pears cost?

11. Mike drove his car 150 miles in 3 hours. What was his average speed in
₍₆₀₎ miles per hour?

12. $46.00
_(28, 52) − $45.56

13.   10,165
₍₅₂₎ −   856

14. $ 0.63
_(11, 28, 51) $ 1.49
       $12.24
       $ 0.38
       $ 0.06
       $ 5.00
   + $ 1.20

15. 70^2
_(65, 85)

16. 71×69
₍₈₆₎

17. $4\overline{)\$30.00}$
₍₈₃₎

18. $3\overline{)263}$
₍₇₂₎

19. $5x = 4080$
_(34, 79)

20. $\dfrac{344}{8}$
₍₆₉₎

21. 37×60
(71)

22. 56×42
(88)

23. $\$5.97 \times 8$
(59)

24. $10.00 - (4.46 - 2.3)$
(45)

*25. Find the median, mode, and range of this set of numbers:
(95)

$$3, 1, 4, 1, 6$$

*26. (Represent) Draw and shade circles to show that 2 equals $\frac{4}{2}$.
(89)

*27. a. (Represent) Draw a square with sides 4 cm long.
(18, 66)

  b. Find the perimeter and the area of the square you drew.

*28. Graph the equation $x = 3$. Name two points on the line.
(Inv. 9)

*29. If $y = 6x - 4$, what is y when
(64, 93)

  a. x is 5?   **b.** x is 8?

*30. (Analyze) In this pattern of loose tiles, there are triangles and squares.
(67)

  a. How many same-sized triangles as the shaded triangle, are there?

  b. What fractional part of the whole design is the shaded triangle?

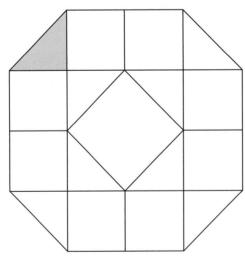

🖊 *California Mathematics Content Standards*

MG 3.0, 3.6 Visualize, describe, and make models of geometric solids (e.g., prisms, pyramids) in terms of the number and shape of faces, edges, and vertices; interpret two-dimensional representations of three-dimensional objects; and draw patterns (of faces) for a solid that, when cut and folded, will make a model of the solid.

• Geometric Solids

facts   Power Up H

mental math

Find half of a product in problems **a–c.**

a. Number Sense: half of 100 × 12

b. Number Sense: half of 100 × 24

c. Number Sense: half of 100 × 48

d. Money: The salad cost $4.89. Ramona paid for it with a $10 bill. How much change should she receive?

e. Geometry: The angles of the triangle measured 47°, 43°, and 90°. What is the sum of the angle measures?

f. Estimation: In 10 minutes, Tevin counted 25 cars that drove through the intersection. About how many cars might Tevin expect to count in 20 minutes?

g. Calculation: 16 ÷ 2, − 6, × 2, $\sqrt{}$

h. Simplify: 1 + 5 × 3

problem solving

There are three light switches that each control a row of lights in the classroom—a row in front, a row in the middle, and a row in back. Make a tree diagram to find the different ways the rows of lights can be turned on or off. Use the tree diagram to count the total number of combinations.

New Concept

Figures such as triangles, rectangles, and circles are flat shapes that cover an area but do not take up space. They have length and width but not depth. Objects that take up space are things such as cars, basketballs, desks, and houses. People also take up space.

Geometric shapes that take up space are called **geometric solids.** The chart below shows the names of some geometric solids.

Geometric Solids

Shape	Name
	Cube and rectangular prism
	Rectangular prism (or rectangular solid)
	Triangular prism
	Pyramid
	Cylinder
	Sphere
	Cone

Example 1

Name each shape:

a.

b.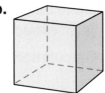

c.

We compare each shape with the chart.

a. sphere   **b. cube**   **c. cone**

Example 2

What is the shape of a soup can?

A soup can has the shape of a **cylinder.**

A flat surface of a solid is called a **face.** Two faces meet at an **edge.** Three or more edges meet at a corner called a **vertex** (plural: *vertices*).

face

edge

vertex

A circular cylinder has one curved surface and two flat circular surfaces. A cone has one curved surface and one flat circular surface. The pointed end of a cone is its **apex.** A sphere has no flat surfaces.

 **Example 3**

> **a. How many faces does a box have?**
>
> **b. How many vertices does a box have?**
>
> **c. How many edges does a box have?**
>
> Find a closed, rectangular box in the classroom (a tissue box, for example) to answer the questions.
>
> **a. 6 faces** (top, bottom, left, right, front, back)
>
> **b. 8 vertices** (4 around the top, 4 around the bottom)
>
> **c. 12 edges** (4 around the top, 4 around the bottom, and 4 running from top to bottom)

Activity

Geometric Solids in the Real World

Material needed:

- **Lesson Activity 27**

Looking around us, we see examples of the geometric solids shown in the table of this lesson. With some objects, two or more shapes are combined. For example, in a building we might see a triangular prism and a rectangular prism.

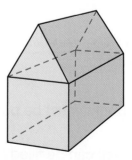

Complete **Lesson Activity 27** by finding and naming an object for each shape. Then draw a picture of each object on the page.

Lesson Practice In problems **a–d,** name the shape of the object listed:

a. basketball

b. shoebox

c. funnel

d. juice can

e. The figure at right is the same shape as several Egyptian landmarks. What is the shape?

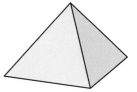

f. The figure in problem **e** has a square base. How many faces does the figure have? How many edges? How many vertices?

Use the rectangular prism to answer problems **g–h.**

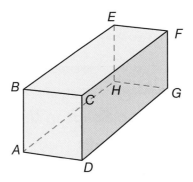

g. Name the front face and the back face. Are the two faces congruent?

h. Name two faces that share edge *CD.* Are the two faces perpendicular?

Written Practice *Distributed and Integrated*

＊**1.** **(Analyze)** All 110 books must be packed in boxes. Each box will hold
(87)   8 books.

   a. How many boxes can be filled?

   b. How many boxes are needed to hold all the books?

*2. **Formulate** What number is five more than the product of six and
(63) seven? Write an expression.

*3. **Explain** Sergio paid $7 for the tape. He received a quarter and
(93) two dimes as change. Tax was 42¢. What was the price of the tape?
Explain how you found your answer.

*4. **Represent** Four fifths of the 600 gymnasts did back handsprings.
(94) How many gymnasts did back handsprings? Draw a picture to
illustrate the problem.

*5. **Explain** Mrs. Tyrone is arranging 29 desks into rows. If she starts
(87) by putting 8 desks in each row, how many desks will be in the last row?
Explain how you know.

6. **Analyze** What is the value of two $100 bills, five $10 bills, four $1
(28) bills, 3 dimes, and 1 penny?

7. a. Find the length of the line segment in millimeters.
(42)

 b. Find the length of the line segment in centimeters. Write the answer
 as a decimal number.

8. **Represent** Use words to write 12.67.
(Inv. 4)

*9. a. Round 3834 to the nearest thousand.
(40, 46)

 b. Round 38.34 to the nearest whole number.

10. The diameter of a circle is 1 meter. What is the radius of the circle in
(18, 42) centimeters?

*11. Find the sum of two hundred eighty-six thousand, five hundred fourteen
(27, 51) and one hundred thirty-seven thousand, two.

12. Seven pairs of beach sandals cost $56. What is the cost of one pair?
(93) What is the cost of ten pairs?

* **13.** (**Interpret**) Find the median, mode, range, and any outliers of
(95) this data:

$$1, 32, 44, 28, 50, 28$$

* **14.** This triangular prism has a rectangular base. How many vertices does
(96) it have?

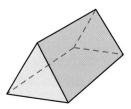

15. $7.48 - (6.47 + 0.5)$
(45)

16. 40×50
(85)

17. 41×49
(86)

18. $2^3 \times 5 \times \sqrt{49}$
(Inv. 3, 65)

19. $\begin{array}{r} 32 \\ \times\ 17 \\ \hline \end{array}$
(88)

20. $\begin{array}{r} 38 \\ \times\ 40 \\ \hline \end{array}$
(71)

21. $7 + 4 + 6 + 8 + 5 + 2 + 7 + 3 + k = 47$
(4)

22. $8\overline{)360}$
(69)

23. $4\overline{)810}$
(83)

24. $7\overline{)356}$
(83)

25. $6n = \$4.38$
(34, 79)

26. $7162 \div 9$
(79)

27. $\dfrac{1414}{2}$
(83)

* **28.** Draw and shade circles to show that 2 equals $\frac{8}{8}$.
(89)

* **29.** The basketball player was 211 centimeters tall. Write the height of the
(42) basketball player in meters.

30. How many square yards of carpeting are needed to cover the floor of a
(66, 84) classroom that is 15 yards long and 10 yards wide?

LESSON 97

Constructing Prisms

California Mathematics Content Standards

MG 3.0, 3.5 Know the definitions of a right angle, an acute angle, and an obtuse angle. Understand that 90°, 180°, 270°, and 360° are associated, respectively, with $\frac{1}{4}$, $\frac{1}{2}$, $\frac{3}{4}$, and full turns.

MG 3.0, 3.6 Visualize, describe, and make models of geometric solids (e.g., prisms, pyramids) in terms of the number and shape of faces, edges, and vertices; interpret two-dimensional representations of three-dimensional objects; and draw patterns (of faces) for a solid that, when cut and folded, will make a model of the solid.

facts   Power Up H

mental math   Fifty is half of 100. Find each product by multiplying by half of 100 in problems **a–d.**

   a. Number Sense: 50×16

   b. Number Sense: 50×44

   c. Number Sense: 50×26

   d. Number Sense: 50×68

   e. Money: The groceries cost $32.48 and the magazine cost $4.99. What was the total cost?

   f. Estimation: Each box is 30.5 cm tall. Estimate the height (using cm) of a stack of 6 boxes.

   g. Calculation: $200 \div 2, \div 2, \div 2$

   h. Simplify: $20 - 10 \div 5$

problem solving   Choose an appropriate problem-solving strategy to solve this problem. There are 365 days in a common year, which is about 52 weeks. However, since 52 weeks is exactly 364 days, a year does not start on the same day of the week as the start of the preceding year. If a common year starts on a Tuesday, on what day of the week will the following year begin?

New Concept

In Lesson 96 we named solids by their shapes. In this lesson we will focus our attention on understanding rectangular prisms and triangular prisms.

Consider the shape of a cereal box. The shape is called a rectangular prism (or rectangular solid). Every panel (side) of a closed cereal box is a rectangle.

If an empty cereal box or similar container is available, you may refer to it to answer the following questions:

1. A closed cereal box has how many panels?

2. What words could we use to refer to these panels?

3. Without a mirror, what is the largest number of panels that can be seen at one time?

4. Two panels meet at a fold, or seam, in the cardboard. Each fold is an edge. A closed cereal box has how many edges?

5. Three edges meet at each corner, or vertex, of the box. A closed cereal box has how many vertices?

If we tape an empty cereal box closed and cut it along seven edges, we can "flatten out" the container, as shown below.

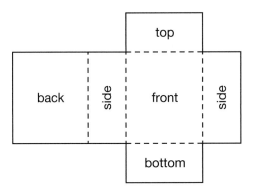

We can see the six rectangles that formed the panels of the closed box. We will use nets like this one to construct the models of solids.

✂ Activity

Constructing Prisms

Materials needed:
- **Lesson Activities 28, 29,** and **30**
- scissors
- tape or glue

| **Math Language** |
| A **net** is a 2-dimensional representation of a 3-dimensional geometric figure. |

We can make models of cubes, rectangular prisms, and triangular prisms by cutting, folding, and taping nets of shapes to form 3-dimensional figures. Use **Lesson Activities 28, 29,** and **30** to construct two rectangular prisms, two triangular prisms, and two cubes. Then study those figures to answer the questions in Lesson Practice.

Lesson Practice   (**Discuss**) Refer to the cube to answer problems **a–e.**

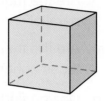

  a. What is the shape of each face?

  b. Is each face parallel to an opposite face?

  c. Is each edge parallel to at least one other edge?

  d. Is each edge perpendicular to at least one other edge?

  e. What type of angle is formed by every pair of intersecting edges?

Refer to the rectangular prism below to answer problems **f–j.**

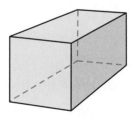

  f. What is the shape of each face?

  g. Is each face parallel to the opposite face?

  h. Is each edge parallel to at least one other edge?

  i. Is each edge perpendicular to at least one other edge?

  j. What type of angle is formed by every pair of intersecting edges?

Refer to the triangular prism with two faces that are equilateral triangles to answer problems **k–o.**

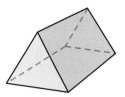

k. What are the shapes of the five faces?

l. Are the triangular faces parallel? Are the rectangular faces parallel?

m. Are the triangular faces congruent? Are the rectangular faces congruent?

n. Do you find pairs of edges that are parallel? That are perpendicular? That intersect but are not perpendicular?

o. What types of angles are formed by the intersecting edges?

Refer to the triangular prism with two faces that are right triangles to answer problems **p–t.**

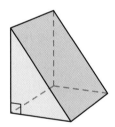

p. What are the shapes of the five faces?

q. Which faces are parallel?

r. Are the triangular faces congruent? Are the rectangular faces congruent?

s. Are there pairs of edges that are parallel? Perpendicular? Intersecting but not perpendicular?

t. What types of angles are formed by the intersecting edges?

u. (**Verify**) One of these nets could be cut out and folded to form a cube. The other will not form a cube. Which net will form a cube?

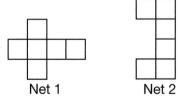

Net 1   Net 2

v. Model Use 1-inch grid paper to draw a different net that will fold into a cube. Cut it out and fold it up to check your work.

Written Practice
Distributed and Integrated

1. Analyze Find an even number between 79 and 89 that can be divided by 6 without a remainder.
(RF6, 68)

2. How many minutes is 3 hours?
(13, 37)

*3. Victor has $8. Dana has $2 less than Victor. How much money do they have altogether?
(93)

4. Represent Write each fraction or mixed number as a decimal number:
(Inv. 4)

   a. $\dfrac{3}{10}$ **b.** $4\dfrac{99}{100}$ **c.** $12\dfrac{1}{100}$

*5. Represent Five eighths of the 40 students wore school colors. How many students wore school colors? Draw a picture to illustrate the problem.
(94)

6. a. What is the diameter of this circle in centimeters?
(18, 42)

   b. What is the radius of this circle in centimeters?

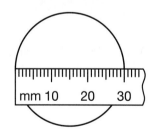

7. The radius of a circle is what fractional part of the diameter?
(18)

8. Estimate the product of 49 and 68. Then find the actual product.
(91)

*9. Explain Pavan has filled a pitcher with iced tea for two guests and himself. The capacity of the pitcher is two quarts. How many 10-ounce glasses of iced tea can be poured from the pitcher? Explain your answer.
(87)

*10. **Estimate** Which letter represents the number closest to 3.5?
(92)

*11. Gretchen paid $20 for five identical bottles of fruit juice. She received
(93) $6 in change. What was the price of one bottle of juice?

*12. **Analyze** Find the median, mode, and range of Vonda's game scores.
(95) (Since there is an even number of scores, the median is the average of
the two middle scores.)

$$100, 80, 90, 85, 100, 90, 100, 100$$

13. $3.85
(59)
$\times \quad 7$

14. 48
(88)
$\times 29$

15.
(11)
16
15
23
8
217
20
6
+ 317

16.
(4)
5
4
3
7
2
5
8
1
4
+ n
45

17. 60^2
(65, 85)

18. 59×61
(88)

19. $\dfrac{400}{5}$
(75)

20. $6\overline{)582}$
(69)

21. $9\overline{)\$37.53}$
(79)

22. $7\overline{)420}$
(83)

23. $7.50 - (3.25 - 0.12)$
(9, 45)

*24. **Represent** Draw and shade circles to show that $3\frac{3}{4}$ equals $\frac{15}{4}$.
(89)

25. The perimeter of this square is 20 inches. What is the
(20, 66) length of each side of the square? What is the area of
the square?

*26. Write a fraction equal to 1 with a denominator of 8.
(89)

*27. Which of the following nets folds up into a triangular prism?
₍₉₇₎

A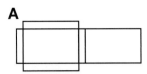

B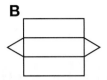

C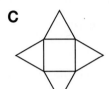

D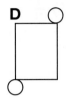

*28. Songhi measured the paper in her notebook and found that it was
₍₄₂₎ 28 cm long. Write the length of her paper in meters.

*29. (**Estimate**) Round $12\frac{5}{12}$ to the nearest whole number.
₍₉₂₎

*30. a. (**Classify**) What is the geometric name for the shape
₍₉₆₎ of a cereal box?

b. How many edges does this box have?

c. Describe the angles.

California Mathematics Content Standards

NS 1.0, **1.4** Decide when a rounded solution is called for and explain why such a solution may be appropriate.

NS 1.0, 1.7 Write the fraction represented by a drawing of parts of a figure; represent a given fraction by using drawings; and relate a fraction to a simple decimal on a number line.

NS 1.0, **1.9** Identify on a number line the relative position of positive fractions, positive mixed numbers, and positive decimals to two decimal places.

• Fractions Equal to 1 and Fractions Equal to $\frac{1}{2}$

facts   Power Up H

mental math   We can double one factor of a multiplication and take one half of the other factor to find a product.

$$4 \times 18$$

double $\downarrow$ $\quad$ $\downarrow$ half

$$8 \times 9 = 72$$

Find each product by the "double and half" method in problems **a–d.**

a. Number Sense: 3×14

b. Number Sense: 4×16

c. Number Sense: 5×22

d. Number Sense: 50×24

e. Money: $\$1.00 - 42¢$

f. Estimation: Choose the more reasonable estimate for the height of a ceiling: 250 cm or 250 m.

g. Calculation: $6^2, + 4, - 30, \times 10$

h. Simplify: $25 - 5 \times 5$

problem solving   Choose an appropriate problem-solving strategy to solve this problem. To get to the room where he will have his yearly medical checkup, Jerome will walk through three doors—a door into the doctor's office building, a door into the waiting room, and a door into the checkup room. Each door might be either open or closed when Jerome gets to it. List the possible combinations of open and closed doors that Jerome might encounter on his way into the checkup room. Use the abbreviations O for "open" and C for "closed."

Each of the following circles is divided into parts. Together, the parts of each circle make up a whole.

We see that 2 halves is the same as 1 whole. We also see that 3 thirds, 4 fourths, and 5 fifths are ways to say 1 whole. If the numerator (top number) and the denominator (bottom number) of a fraction are the same, the fraction equals 1.

$$1 = \frac{2}{2} \qquad 1 = \frac{3}{3} \qquad 1 = \frac{4}{4} \qquad 1 = \frac{5}{5}$$

Example 1

Which of these fractions equals 1?

$$\frac{1}{6} \qquad \frac{5}{6} \qquad \frac{6}{6} \qquad \frac{7}{6}$$

A fraction equals 1 if its numerator and denominator are equal. The fraction equal to 1 is $\frac{6}{6}$.

Model Use fraction manipulatives to verify that $\frac{6}{6} = 1$.

Example 2

Write a fraction equal to 1 that has a denominator of 7.

A fraction equals 1 if its numerator and denominator are the same. If the denominator is 7, the numerator must also be 7. We write $\frac{7}{7}$.

If the numerator of a fraction is half the denominator, then the fraction equals $\frac{1}{2}$. Notice below that the top number of each fraction illustrated is half of the bottom number of the fraction.

$$\frac{1}{2} \qquad \frac{2}{4} \qquad \frac{3}{6} \qquad \frac{4}{8}$$

If the numerator is less than half the denominator, the fraction is less than $\frac{1}{2}$. If the numerator is greater than half the denominator, the fraction is greater than $\frac{1}{2}$.

Model Use fraction manipulatives to verify that $\frac{5}{10} = \frac{1}{2}$.

Example 3

a. **Which fraction below equals $\frac{1}{2}$?**

b. **Which is less than $\frac{1}{2}$?**

c. **Which is greater than $\frac{1}{2}$?**

$$\frac{3}{7} \qquad \frac{6}{12} \qquad \frac{5}{9}$$

a. Since 6 is half of 12, the fraction equal to $\frac{1}{2}$ is $\frac{6}{12}$.

b. Since 3 is less than half of 7, the fraction less than $\frac{1}{2}$ is $\frac{3}{7}$.

c. Since 5 is greater than half of 9, the fraction greater than $\frac{1}{2}$ is $\frac{5}{9}$.

Example 4

Compare: $\frac{3}{8} \bigcirc \frac{1}{2}$

Since 3 is less than half of 8, we know that $\frac{3}{8}$ is less than $\frac{1}{2}$.

$$\frac{3}{8} < \frac{1}{2}$$

Represent Make a sketch that proves the answer is correct.

Example 5

Round $6\frac{7}{10}$ to the nearest whole number.

Halfway between 6 and 7 is $6\frac{1}{2}$. We know that $6\frac{7}{10}$ is greater than $6\frac{1}{2}$ because $\frac{7}{10}$ is greater than $\frac{5}{10}$, which equals $\frac{1}{2}$.

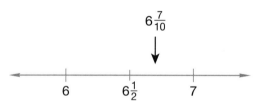

This means $6\frac{7}{10}$ rounds to **7.**

Example 6

Estimate the perimeter and area of this rectangle.

$7\frac{7}{8}$ in.

$4\frac{1}{4}$ in.

First we round each dimension to the nearest whole number of inches. Since $\frac{7}{8}$ is greater than $\frac{1}{2}$, we round $7\frac{7}{8}$ in. up to 8 in. Since $\frac{1}{4}$ is less than $\frac{1}{2}$, we round $4\frac{1}{4}$ in. down to 4 in. Then we use 8 in. and 4 in. to estimate the perimeter and area.

Perimeter: 8 in. + 4 in. + 8 in. + 4 in. = **24 in.**

Area: 8 in. × 4 in. = **32 sq. in.**

Lesson Practice

a. Write a fraction equal to 1 that has a denominator of 6.

b. Multiple Choice Which of these fractions equals 1?

A $\frac{1}{10}$ **B** $\frac{9}{10}$ **C** $\frac{10}{10}$ **D** $\frac{11}{10}$

What fraction name for 1 is shown by each picture?

c.

d.

e. Write a fraction equal to $\frac{1}{2}$ with a denominator of 12.

f. Compare: $\frac{9}{20} \bigcirc \frac{1}{2}$

g. Estimate Round $5\frac{3}{8}$ to the nearest whole number.

h. Estimate the perimeter and area of a rectangle that is $6\frac{3}{4}$ in. long and $4\frac{3}{8}$ in. wide.

Written Practice   *Distributed and Integrated*

*1. a. Analyze If the perimeter of a square is 280 feet, how long is each
(20, 31, 66) side of the square?

b. What is the area?

c. Give the dimensions of another rectangle that has the same perimeter, 280 ft, and an area greater than 100 ft² and less than 150 ft².

*2. There are 365 days in a common year. How many full weeks are there
(RF12, 87) in 365 days?

*3. Nia passed out crayons to 6 of her friends. Each friend received
(87, 93) 3 crayons. There were 2 crayons left for Nia. How many crayons
did Nia have when she began?

*4. (**Represent**) Three fifths of the 60 trees in the orchard were more than
(94) 10 feet tall. How many trees were more than 10 feet tall? Draw a
picture to illustrate the problem.

5. a. Find the length of this line segment in millimeters.
(42)

    b. Find the length of the line segment in centimeters. Write the answer
as a decimal number.

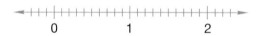

*6. What fraction name for 1 is shown by this circle?
(98)

*7. Round $350,454 to the nearest thousand, to the nearest hundred, and
(22, 40) to the nearest ten.

*8. Copy this number line. Then place a dot at $\frac{1}{2}$ and label the dot point *A.*
(Inv. 4, 43, 92) Place a dot at 1.3 and label the dot point *B.* Place a dot at $1\frac{7}{10}$ and label
the dot point *C.*

$$\overset{\displaystyle \longleftarrow \!+\!+\!+\!|\!+\!+\!+\!+\!+\!+\!+\!+\!+\!|\!+\!+\!+\!+\!+\!+\!+\!+\!|\!+\!+\!+\!\longrightarrow}{\underset{\textstyle 0 \qquad\qquad 1 \qquad\qquad 2}{}}$$

*9. (**Represent**) Change the improper fraction $\frac{5}{4}$ to a mixed number. Draw
(89) a picture to show that the improper fraction and the mixed number are
equal.

10. 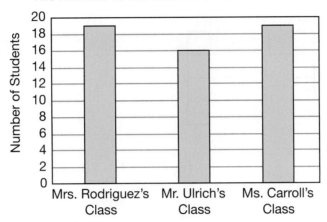 **Interpret** The bar graph shows the number of students in fourth
(95) grade at Sebastian's school. Use the graph to answer the questions
that follow.

The Number of 4th Graders at Sebastian's School

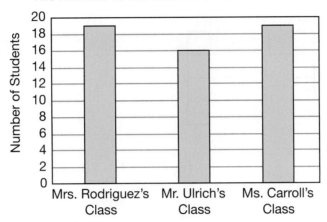

a. How many fewer students are in Mr. Ulrich's class than in
Ms. Carroll's class or in Mrs. Rodriguez's class?

b. Altogether, how many fourth grade students does the bar graph
represent?

c. Which measure of the data is greater: the range or the median?
Explain your answer.

11. The baker used 30 pounds of flour each day to make bread. How many
(39, 71) pounds of flour did the baker use in 73 days?

12. The chef used 132 pounds of potatoes every 6 days. How many
(53, 69) pounds of potatoes were used each day?

* **13.** **Interpret** Jeremy asked 8 friends what type of dog is their favorite.
(95) The results of his survey are below. What is the mode of this data?

German Shepard	Poodle
Boston Terrier	Golden Retriever
Irish Setter	Boston Terrier
Dachshund	Poodle

* 14. a. What is another name for this rectangular prism?
(96, 97)

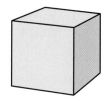

b. How many edges does it have?

c. Which of the following nets will fold up to make the figure?

Net 1   Net 2

15. $80 − ($63.72 + $2)
(9, 28,
52)

16. 37,614 − 29,148
(52)

17. 9w = 9 · 26
(64)

18. 3⁴
(65)

18. 3^4
(65)

19. 24 × 1000
(84)

20. 79¢ × 6
(28, 38)

21.   50
(85)   × 50

22.   51
(86)   × 49

23.   47
(88)   × 63

24. 4)810
(79)

25. 5)490
(69)

26 6)362
(72)

27. 1435 ÷ $\sqrt{49}$
(Inv. 3,
83)

* 28. a. ⟨ **Analyze** ⟩ How many 8-ounce cups of milk can be poured from
(73)   one gallon of milk?

b. Write a formula to change any number of gallons to cups. Use *g* for
gallons and *c* for cups.

* 29. Round $16\frac{5}{8}$ to the nearest whole number.
(98)

* 30. Estimate the area of a window with the dimensions shown:
(98)

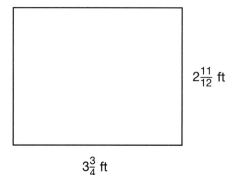

$2\frac{11}{12}$ ft

$3\frac{3}{4}$ ft

Changing Improper Fractions to Whole Numbers or Mixed Numbers

✎ *California Mathematics Content Standards*

NS 1.0, 1.5 Explain different interpretations of fractions, for example, parts of a whole, parts of a set, and division of whole numbers by whole numbers; explain equivalents of fractions (see Standard 4.0).

NS 1.0, 1.6 Write tenths and hundredths in decimal and fraction notations and know the fraction and decimal equivalents for halves and fourths (e.g., $\frac{1}{2}$ = 0.5 or .50; $\frac{7}{4}$ = $1\frac{3}{4}$ = 1.75).

NS 1.0, 1.7 Write the fraction represented by a drawing of parts of a figure; represent a given fraction by using drawings; and relate a fraction to a simple decimal on a number line.

Power Up

facts   Power Up J

mental math   Find each product by the "double and half" method in problems **a–c**.

  a. Number Sense: 3×18

  b. Number Sense: 15×60

  c. Number Sense: 50×48

  d. Money: Shawntay had $5.00. He spent $1.75 on a birthday card for his brother. How much does he have left?

  e. Fractional Part: What is $\frac{1}{5}$ of 100?

  f. Estimation: Brittany used $11\frac{3}{4}$ inches of tape to wrap one gift. About how much tape will she need to wrap five more gifts that are the same size as the first?

  g. Calculation: $\sqrt{25}$, $\div$ 5, + 6, $\times$ 2, − 11

  h. Simplify: $1 + (2 \times 3)$

problem solving   Choose an appropriate problem-solving strategy to solve this problem. Danae can walk twice as fast as she can swim. She can run twice as fast as she can walk. She can ride a bike twice as fast as she can run. If Danae can ride her bike a quarter mile in one minute, how long would it take her to swim a quarter mile?

New Concept

If the numerator of a fraction is equal to or greater than the denominator, the fraction is an improper fraction. All of these fractions are improper fractions:

$$\frac{3}{2} \qquad \frac{5}{4} \qquad \frac{10}{3} \qquad \frac{9}{4} \qquad \frac{5}{5}$$

Model Use fraction manipulatives to show $\frac{3}{2}$ and $\frac{5}{4}$ as mixed numbers.

To write an improper fraction as a whole or mixed number, we divide to find out how many wholes the improper fraction contains. If there is no remainder, we write the improper fraction as a whole number. If there is a remainder, the remainder becomes the numerator in a mixed number.

Example 1

Write $\frac{13}{5}$ as a mixed number. Draw a picture to show that the improper fraction and mixed number are equal.

To find the number of wholes, we divide.

$$\begin{array}{r} 2 \leftarrow \text{wholes} \\ 5{\overline{\smash{\big)}\,13}} \\ \underline{10} \\ 3 \leftarrow \text{remainder of 3} \end{array}$$

This division tells us that $\frac{13}{5}$ equals two wholes with three fifths left over. We write this as **$2\frac{3}{5}$**. We can see that $\frac{13}{5}$ equals $2\frac{3}{5}$ if we draw a picture.

$$\frac{13}{5} \quad = \quad \frac{5}{5} \quad + \quad \frac{5}{5} \quad + \quad \frac{3}{5} \quad = \quad 2\frac{3}{5}$$

Example 2

Write $\frac{10}{3}$ as a mixed number. Then draw a picture to show that the improper fraction and mixed number are equal.

First we divide.

$$\begin{array}{r} 3 \\ 3{\overline{\smash{\big)}\,10}} \\ \underline{9} \\ 1 \end{array}$$

From the division we see that there are three wholes. One third is left over. We write **$3\frac{1}{3}$**. Then we draw a picture to show that $\frac{10}{3}$ equals $3\frac{1}{3}$.

$\frac{10}{3} = 3\frac{1}{3}$

(**Formulate**) Give a real-world example for dividing items into groups of $3\frac{1}{3}$.

Example 3

Write $\frac{12}{4}$ as a whole number. Then draw a picture to show that the improper fraction and whole number are equal.

First we divide.

$$
\begin{array}{r}
3 \\
4\overline{)12} \\
\underline{12} \\
0
\end{array}
$$

We have three wholes and no remainder. Our picture looks like this:

  　　$\frac{12}{4} = \mathbf{3}$

(**Discuss**) Explain how $\frac{4}{4}$ is related to $\frac{12}{4}$.

Example 4

Write each improper fraction as a decimal.

　a. $\dfrac{7}{4}$　　　　　　　**b.** $\dfrac{17}{10}$

　a. We can change $\frac{7}{4}$ to a mixed number.

$$\frac{7}{4} = \frac{4}{4} + \frac{3}{4} \text{ or } 1\frac{3}{4}$$

We know that $\frac{3}{4}$ is 0.75, so $1\frac{3}{4}$ is **1.75.**

　b. We can write $\frac{17}{10}$ as a mixed number.

$$\frac{17}{10} = \frac{10}{10} + \frac{7}{10} \text{ or } 1\frac{7}{10}$$

We know that $1\frac{7}{10}$ is also written as **1.7.**

Represent Change each improper fraction to a whole number or to a mixed number. Then draw a picture to show that the improper fraction is equal to the number you wrote.

a. $\frac{7}{2}$ **b.** $\frac{12}{3}$ **c.** $\frac{8}{3}$ **d.** $\frac{15}{5}$

Analyze Write each improper fraction as a decimal:

e. $\frac{7}{2}$ **f.** $\frac{19}{10}$

Written Practice

Distributed and Integrated

1. How many 6¢ erasers can be bought with 2 quarters?
(87)

2. Two quarters are what fractional part of a dollar?
(29)

3. Jason has $8. Parisa has $2 more than Jason. How much money do they have altogether?
(93)

*4. **Represent** Three fourths of the 20 students in a class participate in an after-school activity. What number of students participate? Draw a picture to illustrate and solve the problem.
(94)

*5. **Interpret** Bethany surveyed twelve friends and asked them the number of times they saw a movie in a theater last month. She recorded the results on the number line below.
(95)

Number of Movies in a Theater

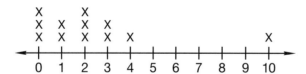

a. **Interpret** Is there an outlier? If yes, name it.

b. **Explain** How can you find the median for this data?

c. **Verify** What is the mode? Explain how you know.

d. What is the range of this data?

*6. Write a fraction equal to one that has a denominator of 10.
(98)

7. (Represent) Write 86.74 with words.
(Inv. 4)

*8. (Estimate) There are many ways to make an estimate. Describe two
(61) different ways to estimate the difference of 496 subtracted from 605.

*9. Change each improper fraction to a whole number or a mixed number:
(99)

    a. $\dfrac{9}{5}$ **b.** $\dfrac{9}{3}$ **c.** $\dfrac{9}{2}$

    d. Write each number in order from greatest to least as improper fractions.

*10. (Estimate) Soon after James Marshall discovered gold at John Sutter's
(61, 76) mill in California on January 24, 1848, the "gold rush" began. If 2450
people came in 5 days, about how many came each day? About how
many people came in 1 week?

11. Find the length of this segment to the nearest tenth of a centimeter.
(42) Write the length as a decimal number.

*12. A miner bought 6 bags of flour for $4.20 per bag and 8 pounds of salt
(93) for 12¢ per pound. How much money did the miner spend?

*13. **a.** Which digit in 86.74 is in the tenths place?
(41, 62)

    b. Is 86.74 closer to 86.7 or 86.8?

14. Draw a trapezoid.
(90)

15. $4.86 - (2.8 + 0.56)$
(9, 45)

16. 30^2 **17.** 54×29
(65, 85)   *(88)*

*18. $5\overline{)230}$ **19.** $7\overline{)2383}$
(69)   *(83)*

*20. Which letters in **MATH** have one line of symmetry? Which has two lines
(82) of symmetry? Which has rotational symmetry?

*21. $372 \div 5$
(72)

22. $8c = \$5.76$
(34, 79)

23.
(11)

$$\begin{array}{r} 12 \\ 26 \\ 13 \\ 35 \\ 110 \\ 8 \\ + \;\; 15 \\ \hline \end{array}$$

24. $351,426$
(51) $+ 449,576$
$\overline{}$

25. $\$50.00$
(52) $- \$49.49$
$\overline{}$

26. $\$12.49$
(59) $\times 8$
$\overline{}$

27. 73
(88) $\times 62$
$\overline{}$

*28. a. A field is 300 feet long and 200 feet wide. How many
(20) feet of fencing would be needed to go around the field?

300 ft

200 ft

b. **Explain** Is this problem about perimeter or area? How do you know?

*29. a. Name a geometric solid that has no vertices.
(96, 97)

b. **Justify** Can a rectangular prism have no congruent faces? Why or why not?

*30. **Interpret** Use this chart to answer parts **a–c.**
(RF13)

Mileage Chart

	Atlanta	Boston	Chicago	Kansas City	Los Angeles	New York City	Wash., D.C.
Chicago	674	963		499	2054	802	671
Dallas	795	1748	917	489	1387	1552	1319
Denver	1398	1949	996	600	1059	1771	1616
Los Angeles	2182	2979	2054	1589		2786	2631
New York City	841	206	802	1198	2786		233
St. Louis	541	1141	289	257	1845	948	793

a. The distance from Los Angeles to Boston is how much greater than the distance from Los Angeles to New York City?

b. Rebecca is planning a trip from Chicago to Dallas to Los Angeles to Chicago. How many miles will her trip be?

c. There are three empty boxes in the chart. What number would go in these boxes?

❧ *California Mathematics Content Standards*

NS 1.0, 1.5 Explain different interpretations of fractions, for example, parts of a whole, parts of a set, and division of whole numbers by whole numbers; explain equivalents of fractions (see Standard 4.0).

NS 1.0, 1.7 Write the fraction represented by a drawing of parts of a figure; represent a given fraction by using drawings; and relate a fraction to a simple decimal on a number line.

MR 2.0, 2.6 Make precise calculations and check the validity of the results from the context of the problem.

• Adding and Subtracting Fractions with Common Denominators

facts	Power Up J
mental math	Find each product by the "double and half" method in problems a–c.

   a. Number Sense: 4×14

   b. Number Sense: 25×80

   c. Number Sense: 50×64

   d. Money: Cooper paid for a lawn sprinkler that cost $8.16 with a $10 bill. How much change should he receive?

   e. Geometry: What is the diameter of a wheel that has a radius of 14 inches?

   f. Estimation: Estimate 19×41 by rounding each number to the nearest ten before multiplying.

   g. Calculation: $15 - 9$, square the number, $\div\, 4$, $-\, 8$

   h. Simplify: $10 - (4 \div 2)$

problem solving

Franklin's family is moving to a new house, and they have packed their belongings in identical boxes. The picture at right represents the stack of boxes that is inside the moving truck. How many boxes are in the stack?

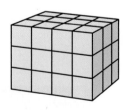

Focus Strategy: Make It Simpler

(**Understand**) We are shown a picture of identical, stacked boxes. We assume that boxes in the upper layer are supported by boxes in the lower layers. We are asked to find how many boxes are in the stack altogether.

Plan In the picture, we can see three layers of boxes. If we can find how many boxes are in each layer, we can multiply by 3 to find the total number of boxes.

Solve If we look at the top layer of boxes, we see 4 boxes along the front and 3 boxes along the side. Four rows of 3 boxes means there are 4 × 3 boxes = 12 boxes in the top layer. The middle and bottom layers contain the same number of boxes as the top layer. Since there are three layers of boxes, we find that there are 3 × 12 boxes = **36 boxes** in the stack altogether.

Check We know our answer is reasonable because three layers of 12 boxes each is 36 boxes altogether. If we have blocks or unit cubes, we can check our answer by modeling the problem.

New Concept

To add fractions, it helps to think of the numerators as objects, like apples. Just as 1 apple plus 1 apple equals 2 apples, 1 third plus 1 third equals 2 thirds.

1 apple	+	1 apple	=	2 apples

1 third	+	1 third	=	2 thirds

$$\frac{1}{3} \quad + \quad \frac{1}{3} \quad = \quad \frac{2}{3}$$

When we add fractions, we add the numerators (top numbers). We do not add the denominators (bottom numbers).

Example 1

Blake mixed $\frac{3}{5}$ of a pound of cashews with $\frac{1}{5}$ of a pound of pecans. What is the weight in pounds of the cashew and pecan mixture?

We add only the top numbers. Three fifths plus one fifth is four fifths. The weight of the cashew and pecan mixture is $\frac{4}{5}$ **of a pound.**

$$\frac{3}{5} + \frac{1}{5} = \frac{4}{5}$$

Likewise, when we subtract fractions, we subtract only the numerators. The denominator does not change. For example, five sevenths minus two sevenths is three sevenths.

$$\frac{5}{7} - \frac{2}{7} = \frac{3}{7}$$

Example 2

To make a small bow for a present, Sakura cut $\frac{1}{5}$ of a yard of ribbon from a length of ribbon that was $\frac{3}{5}$ of a yard long. What is the length of the ribbon that was not used for the bow?

We subtract only the numerators. Three fifths minus one fifth is two fifths. The length of the ribbon not used for the bow is $\frac{2}{5}$ **of a yard.**

$$\frac{3}{5} - \frac{1}{5} = \frac{2}{5}$$

Discuss How can we check the answer?

Recall that a mixed number is a whole number plus a fraction, such as $2\frac{3}{5}$. To add mixed numbers, we add the fraction parts and then the whole-number parts.

Example 3

Add: $2\frac{3}{5} + 3\frac{1}{5}$

It is helpful to write the numbers one above the other. First we add the fractions and get $\frac{4}{5}$. Then we add the whole numbers and get 5. The sum of the mixed numbers is $5\frac{4}{5}$.

$$\begin{array}{r} 2\frac{3}{5} \\ + 3\frac{1}{5} \\ \hline 5\frac{4}{5} \end{array}$$

Example 4

Subtract: $5\frac{2}{3} - 1\frac{1}{3}$

We subtract the second number from the first number. To do this, we write the first number above the second number. We subtract the fractions and get $\frac{1}{3}$. Then we subtract the whole numbers and get 4. The difference is $4\frac{1}{3}$.

$$\begin{array}{r} 5\frac{2}{3} \\ - 1\frac{1}{3} \\ \hline 4\frac{1}{3} \end{array}$$

Example 5

In the race Martin rode his bike $7\frac{1}{2}$ miles and ran $2\frac{1}{2}$ miles. Altogether, how far did Martin ride his bike and run?

This is a story about combining. We add $7\frac{1}{2}$ miles and $2\frac{1}{2}$ miles. The two half miles combine to make a whole mile. The total distance is **10 miles.**

$$\begin{array}{r} 7\frac{1}{2} \\ + 2\frac{1}{2} \\ \hline 9\frac{2}{2} \end{array} = 10$$

Lesson Practice   Find each sum or difference:

a. $\frac{1}{3} + \frac{1}{3}$

b. $\frac{1}{4} + \frac{2}{4}$

c. $\frac{3}{10} + \frac{4}{10}$

d. $\frac{2}{3} - \frac{1}{3}$

e. $\frac{3}{4} - \frac{2}{4}$

f. $\frac{9}{10} - \frac{6}{10}$

g. $2\frac{1}{4} + 4\frac{2}{4}$

h. $5\frac{3}{8} + 1\frac{2}{8}$

i. $8 + 1\frac{2}{5}$

j. $4\frac{3}{5} - 1\frac{1}{5}$

k. $9\frac{3}{4} - 4\frac{2}{4}$

l. $12\frac{8}{9} - 3\frac{3}{9}$

m. How much is three eighths plus four eighths?

n. The troop hiked to the end of the trail and back. If the trail was $3\frac{1}{2}$ miles long, how far did the troop hike?

Written Practice   *Distributed and Integrated*

---

* 1. Use this information to answer parts **a–c.**
(76, 100)

Nara has 6 cats. Each cat eats $\frac{1}{2}$ can of food each day. Cat food costs 47¢ per can.

a. How many cans of cat food are eaten each day?

b. How much does Nara spend on cat food per day?

c. How much does Nara spend on cat food in a week?

* 2. a. Sketch a right triangle. Label the vertices *A*, *B*, and *C*, so that *C*
(17, 50, 81)   is at the right angle.

b. Name two segments that are perpendicular.

c. Name two segments that intersect but are not perpendicular.

d. Can a triangle have two parallel sides?

3. If the perimeter of a square classroom is 120 feet, then how long
(20, 66, 85)   is each side of the classroom? What is the area of the classroom?

* 4. **Represent**   Math was the favorite class of five sevenths of the
(94)   28 students. Math was the favorite class of how many students? Draw a picture to illustrate the problem.

*5. **(Interpret)** Name the median, mode and range of this data.
(95)

14, 23, 34, 51, 63, 23, 14, 23, 45

*6. **(Conclude)** Use this geometric solid to answer parts **a–c.**
(96, 97)

   a. What is the name of this geometric solid?

   b. Name two faces that share (the edge) $\overline{BC}$.

   c. Name three faces that share the same vertex.

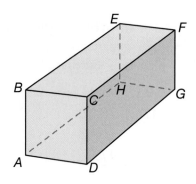

7. If the radius of a circle is $1\frac{1}{2}$ inches, then what is the diameter of the
(18, 100) circle?

8. **(Represent)** Use words to write 523.43.
(Inv. 4)

9. **(Estimate)** Colin used rounding to estimate the product of 61 and 397.
(91) What estimate did Colin make? Explain your answer.

*10. Change each improper fraction to a whole number or a mixed number:
(99)

   a. $\dfrac{10}{10}$ **b.** $\dfrac{10}{5}$ **c.** $\dfrac{10}{3}$

   d. Write each number in order from least to greatest as improper
      fractions.

*11. LaTonya went to the fair with $20. She paid $6.85 for a necklace and
(93) $4.50 for lunch. Then she bought a soft drink for 75¢. How much
money did she have left?

*12. **(Explain)** Clara bought two dolls priced at $7.40 each. The tax was
(93) 98¢. She paid the clerk with a $20 bill. How much change did she get
back? Explain why your answer is reasonable.

13. The big truck that transported the Ferris wheel could go only 140 miles
(60) in 5 hours. What was the truck's average speed in miles per hour?

*14. Compare: $\dfrac{49}{100} \bigcirc \dfrac{1}{2}$
(92, 98)

*15. a. **Estimate** Round $12.25 to the nearest dollar.
(46)

   b. Round 12.25 to the nearest whole number.

*16. a. Which digit in 36.47 is in the tenths place?
(41, 62)

   b. **Estimate** Is 36.47 closer to 36.4 or to 36.5?

17. 73.48
(45)   5.63
   + 17.9

18. $65.00
(28)   − $29.87

19. 24,375
(52)   − 8,416

20. $3.68
(59)   ×   9

21. 89 × 91
(88)

22. 6)3210
(79)

*23. 5)4300
(83)

24. 6)$57.24
(79)

25. 765 ÷ 9
(69)

*26. 563 ÷ 7
(72)

*27. **Evaluate** Find the value of n^2 when n is 90.
(63, 65, 85)

*28. **Evaluate** Find the value of $\frac{m}{\sqrt{m}}$ when m is 36.
(Inv. 3)

*29. a. **Multiple Choice** The sum of $6\frac{3}{4}$ and $5\frac{3}{5}$ is between which two
(98)   numbers?

   A 5 and 7   **B** 30 and 40   **C** 0 and 2   **D** 11 and 13

   b. Explain your answer for part **a.**

30. The African bush elephant is the heaviest land mammal on Earth. Even
(59, 80, 84)   though it eats only twigs, leaves, fruit, and grass, an African bush
   elephant can weigh 7 tons. Seven tons is how many pounds?

California Mathematics Content Standards

MG **2.0**, **2.1** Draw the points corresponding to linear relationships on graph paper (e.g., draw 10 points on the graph of the equation $y = 3x$ and connect them by using a straight line).

MG **2.0**, **2.2** Understand that the length of a horizontal line segment equals the difference of the x-coordinates.

MG **2.0**, **2.3** Understand that the length of a vertical line segment equals the difference of the y-coordinates.

SDAP 1.0, 1.3 Interpret one-and two-variable data graphs to answer questions about a situation.

Focus on

Graphing Relationships

Graphs can also be used to display relationships between two quantities, such as pay and time worked.

Suppose Dina has a job that pays $10 per hour. This table shows the total pay Dina would receive for 1, 2, 3, or 4 hours of work.

1. **Represent** Copy and extend the table to show Dina's pay for each hour up to 8 hours of work.

Pay Schedule

Hours Worked	Total Pay
1	$10
2	$20
3	$30
4	$40

The graph below shows the same relationship between hours worked and total pay. Each dot on the graph represents both a number of hours and an amount of pay.

If Dina works more hours, she earns more pay. We say that her total pay is a function of the number of hours she works. Since Dina's total pay depends on the number of hours she works, we make "Total Pay" the vertical scale and "Total Hours" the horizontal scale.

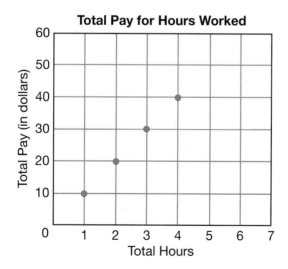

Total Pay for Hours Worked

2. **Represent** Copy the graph. Extend the sides of the graph to include 8 hours and $80. Then graph (draw) the dots for Dina's total pay for each hour up to 8 hours.

The following table and graph show how many miles Rosita hiked at 4 miles per hour:

Miles Hiked
(at 4 mi per hr)

Hours	Miles
1	4
2	8
3	12

Miles Hiked
(at 4 mi per hr)

The dots indicate how far Rosita hiked in one, two, and three hours. However, every second Rosita hiked, she was hiking a small part of a mile. We show this progress by drawing a line through the dots. Every point on a line represents a distance hiked for a given time.

For example, straight up from $1\frac{1}{2}$ hours is a point on the line at 6 miles.

3. **Interpret** Use the graph to find the distance Rosita hiked in $2\frac{1}{2}$ hours.

4. **Analyze** What multiplication formula could you write to represent the relationship between the two sets of data?

5. **Verify** Use your formula to find the number of miles Rosita would hike in 5 hours.

Activity

Graphing Pay Rates

Formulate Work with a partner and agree on an hourly rate of pay for a selected job. Then create a table to display a pay schedule showing the total pay for 1, 2, 3, 4, 5, 6, 7, and 8 hours of work at the agreed rate of pay. Use the pay schedule to create a graph that shows the relationship represented by the table. Write an equation to represent the data.

We can represent similar relationships using equations and graphs.

Suppose a rug cleaning machine can be rented for $2 per hour plus a $10 rental fee. How much would it cost to rent the machine for 3 hours? The hourly rate will be applied to part of an hour.

We can write an equation to represent this problem.

$2 per hour + $10 rental fee = total cost

$$2h + 10 = t$$

We can create a set of ordered pairs for 1, 2, and 3 hours.

$2h + 10 = t$

h	t
1	12
2	14
3	16

Now we can graph the ordered pairs.

Rug Cleaner Rental

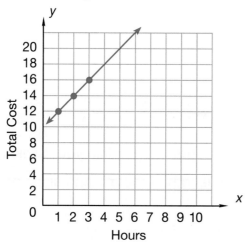

6. **Interpret** Use the graph to find the total cost of renting the machine for $5\frac{1}{2}$ hours.

7. **Analyze** Use the equation to find the cost of renting the machine for 1 full day.

8. Use this problem to answer parts **a–c.**

Mrs. Becker hired a student clown for a children's party. The clown charges $4 per hour plus a $2 traveling fee. The hourly rate is applied to part of an hour.

   a. **Formulate** Write an equation to represent the relationship between hours and pay for any number of hours.

   b. **Evaluate** Use the equation to make a set of ordered pairs for 1, 2, and 3 hours.

   c. **Model** Graph the ordered pairs and draw a line to represent the equation.

The two segments below represent opposite sides of a rectangle:

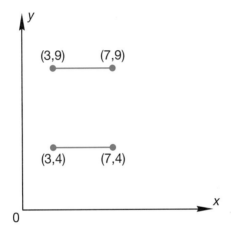

9. **Interpret** What is the length of each segment? Explain your thinking.

10. **Explain** How can you determine the length of the missing sides of the rectangle?

11. **Analyze** What is the perimeter and area of the rectangle?

Investigate Further

   a. How can you subtract coordinates to determine the perimeter of this equilateral triangle?

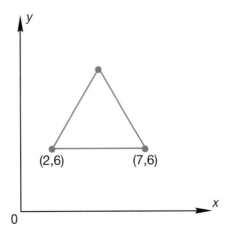

California Mathematics Content Standards

NS 1.0, 1.8 Use concepts of negative numbers (e.g., on a number line, in counting, in temperature, in "owing").

AF 1.0, 1.4 Use and interpret formulas (e.g., area = length × width or A = lw) to answer questions about quantities and their relationships.

MR 2.0, 2.2 Apply strategies and results from simpler problems to more complex problems.

• Formulas

facts   Power Up J

mental math

Thinking of quarters can make mentally adding and subtracting numbers ending in 25, 50, and 75 easier. Use this strategy to solve problems **a–c.**

a. Number Sense: 350 + 175

b. Number Sense: 325 − 150

c. Number Sense: 175 + 125

d. Money: Each ticket costs $10.00 if purchased at the concert hall. A ticket costs $1.95 less if it is purchased in advance. What is the advance price for a ticket?

e. Time: The year 2011 begins on a Saturday. On what day of the week will the year 2012 begin?

f. Estimation: Estimate 24 × 21. Round 24 to 25, round 21 to 20, and then multiply.

g. Calculation: $\frac{1}{10}$ of 70, − 5, × 50, $\sqrt{}$

h. Simplify: 3 × (1 + 1)

problem solving

Choose an appropriate problem-solving strategy to solve this problem. Congress meets in Washington, D.C., to make laws for the United States. The 535 members of the U.S. Congress are divided into two groups—representatives and senators. There are 2 senators from each of the 50 states. The rest of the people in the U.S. Congress are representatives. How many senators are there? How many representatives are there?

We have been using formulas to find area and perimeter.

$$A = lw \qquad P = 2l + 2w$$

We also have been writing formulas to convert from one unit of measure to another.

Feet and Yards	Quarts and Gallons
$f = 3y$	$q = 4g$

Any situation where the relationship between two quantities is constant can be represented with a formula. We can solve problems using these formulas:

Total Cost (T) = Original Price (P) + Sales Tax (S)

Profit (or Loss) (P) = Sales (S) − Expenses (E)

Total Miles (T) = Total Gallons of Fuel (G) × Miles per Gallon (M)

Distance (D) = Speed (S) × Time (T)

Example 1

A hat cost $27.90. The tax is $1.40. What is the total cost of the hat?

We can solve any sales tax problem using this formula:

Total Cost (T) = Original Price (P) + Sales Tax (S) or $T = P + S$

$$T = \$27.90 + \$1.40$$
$$= \$29.30$$

The total cost of the hat is **$29.30.**

Example 2

Lisa spent $25 on supplies to make 20 bracelets. She sold them for $35. What was her profit?

We can use the formula

Profit (P) = Sales (S) − Expenses (E) or $P = S - E$

$$P = \$35 - \$25$$
$$= \$10$$

Lisa's profit was **$10.**

Connect Suppose Lisa sold the bracelets for $20. What would her profit be?

Example 3

Ed's car holds 20 gallons of gasoline. The car dealer told him he could drive 17 miles on each gallon of gas. How far can Ed drive on a full tank?

For this problem we use the formula

Total Miles (T) = Total Gallons of Fuel (G) × Miles per Gallon (M) or $T = G \times M$

$$T = 20 \times 17$$
$$= 340$$

Ed can drive **340 miles** on a full tank.

Lesson Practice

Generalize Write a formula to solve each problem:

a. Genevieve drove for 4 hours at a speed of 50 miles per hour. How far did she drive?

b. At the end of a week, one store had expenses of $1000 and sales of $800. What number would represent the week's record of sales and expenses? Explain why.

c. Mary Beth ran 6 kilometers. How many meters did she run?

d. A car can be driven 95 miles on 5 gallons of gas. How far can it be driven on 1 gallon of gas?

Written Practice
Distributed and Integrated

* 1. **Justify** Haley bought 5 tickets for $2.75 each. She paid for them
(28, 39, 52) with a $20 bill. How much change should she receive? Explain why your answer is reasonable.

2. If fifty cents is divided equally among 3 friends, there will be some
(87) cents left. How many cents will be left?

3. What is the difference when four hundred nine is subtracted from
(10) nine hundred four?

* 4. **Represent** Two fifths of the 45 stamps were from Brazil. How
(94) many stamps were from Brazil? Draw a picture to illustrate the problem.

*5. a. The quesadilla was cut into 10 equal slices. The entire sliced
(19, 98)   quesadilla shows what fraction name for 1?

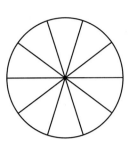

b. Two slices of the quesadilla is what fractional part of the
whole quesadilla?

*6. (**Formulate**) If a car gets 15 miles per gallon, how many miles can the
(58, 101) car go on 15 gallons? Write a formula to solve the problem.

*7. (**Model**) Use grid paper to draw a net that can be folded into a
(97) rectangular prism.

8. (**Estimate**) Round 5167 to the nearest thousand.
(40)

*9. Change the improper fraction $\frac{9}{4}$ to a mixed number.
(99)

*10. Multiple Choice Which of these fractions is *not* equal to 1?
(98)

A $\frac{12}{12}$ **B** $\frac{11}{11}$ **C** $\frac{11}{10}$ **D** $\frac{10}{10}$

11. In the summer of 1926, there were only 17 stores in the town. Today
(38, 39, 76) there are 8 times as many stores in the town. How many stores are in
the town today?

12. (**Estimate**) The wagon train took 9 days to make the 243-mile journey.
(91) About how many miles did it travel per day?

*13. (**Explain**) On Saturday Jacinda played outside for $1\frac{1}{2}$ hours and
(100) played board games for $2\frac{1}{2}$ hours. Altogether, how much time did
Jacinda spend playing outside and playing board games? Explain
how you found your answer.

*14. (**Estimate**) Round $8\frac{21}{100}$ to the nearest whole number.
(98)

15.   36.31
(45) $-$ 7.4

*16. $\frac{5}{8} + \frac{2}{8}$
(100)

17.   6
(4)   5
4
3
$+ n$
25

*18. $\frac{9}{10} - \frac{2}{10}$
(100)

*19. $3\frac{2}{5} + 1\frac{1}{5}$
(100)

20. 27 × 32
(88)

21. 62 × 15
(88)

22. $7^2 + \sqrt{49}$
(Inv. 3, 65)

*__23.__ 5)‾460
(69)

24. 9)‾$27.36
(83)

25. 6w = 2316
(34, 79)

26. 1543 ÷ 7
(83)

*__27.__ 532 ÷ 6
(72)

28. $\dfrac{256}{8}$
(69)

*__29.__ **a.** How many square feet of shingles are needed to cover a rectangular
(Inv. 3, 85)   roof that is 40 feet wide and 60 feet long?

b. Is this problem about area or perimeter? How do you know?

30. Troy walked $2\frac{1}{5}$ miles on Monday. He walked $3\frac{4}{5}$ miles on Wednesday.
(100)   How many more miles did Troy walk on Wednesday than on
Monday?

Real-World Connection

What is the length of the line segment on the grid? Explain how you found your answer.

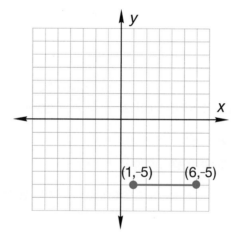

California Mathematics Content Standards
NS **3.0, 3.1** Demonstrate an understanding of, and
the ability to use, standard algorithms for
the addition and subtraction of multidigit
numbers.
NS **3.0, 3.2** Demonstrate an understanding of, and
the ability to use, standard algorithms for
multiplying a multidigit number by a two-
digit number and for dividing a multidigit
number by a one-digit number; use
relationships between them to simplify
computations and to check results.
AF 1.0, 1.2 Interpret and evaluate mathematical
expressions that now use parentheses.

• The Distributive Property

Power Up

facts   Power Up A

**mental
math**

a. Number Sense: 425 − 175

b. Number Sense: 4 × 18

c. Money: Gabriella purchased a sandwich for $3.65 and a beverage for $0.98. What was the total price?

d. Geometry: How many vertices do 4 hexagons have?

e. Time: The year 2012 begins on a Sunday. On what day of the week will the year 2013 begin? (Remember that 2012 is a leap year.)

f. Estimation: Estimate 19 × 31 by rounding one number up and the other number down.

g. Calculation: 4 × 5, − 5, + 6, ÷ 7

h. Simplify: 12 ÷ (6 − 2)

**problem
solving**

Choose an appropriate problem-solving strategy to solve this problem. Nalo said, "An inch is less than $\frac{1}{10}$ of a foot." Write a short paragraph explaining why you agree or disagree with Nalo's statement.

New Concept

The multiplication algorithm is based on the **Distributive Property.** The Distributive Property applies to multiplication problems such as 23 × 14. According to the Distributive Property:

$$23 \times 14 = 23(10 + 4)$$

When we multiply each addend and then add the products, we are using the same approach as the multiplication algorithm.

$$23\,(10 + 4) = (23 \times 10) + (23 \times 4)$$
$$= 230 + 92$$
$$= 322$$

$$\begin{array}{r} 23 \\ \times\ 14 \\ \hline 92 \\ +\ 230 \\ \hline 322 \end{array}$$

This method matches the partial products of the multiplication algorithm.

We can also use the Distributive Property to check a division problem.

$$84 \div 4 = (80 \div 4) + (4 \div 4)$$
$$= 20 + 1$$
$$= 21$$

$$\begin{array}{r} 21 \\ 4\overline{)84} \\ 80 \\ \hline 04 \\ 4 \\ \hline 0 \end{array}$$

This method matches the partial products of the division algorithm.

Example 1

Nina wants to multiply 35 by 25. Use the Distributive Property to find the product.

First we rewrite the factors using parentheses and the expanded form. Then we solve the equation.

$$35(20 + 5) = (35 \times 20) + (35 \times 5)$$
$$= 700 + 175$$
$$= 875$$

The product is **875**.

Example 2

Jose divided 126 by 3. Use the Distributive Property to check his answer.

$$\begin{array}{r} 42 \\ 3\overline{)126} \\ 12 \\ \hline 06 \\ 6 \\ \hline 0 \end{array}$$

$$126 \div 3 = (120 \div 3) + (6 \div 3)$$
$$= 40 + 2$$
$$= 42$$

The quotient **42** is correct.

Connect Use the Distributive Property to find each product.

a. 42×16

b. 38×12

c. **Justify** Divide 245 by 5. Then, use the Distributive Property to check the answer.

Written Practice

Distributed and Integrated

*1. **Explain** Cody bought 8 pounds of oranges. He gave the storekeeper a $5 bill and received $1.96 in change. What did 1 pound of oranges cost? What is the first step in solving this problem?
(93)

2. **Formulate** What number is six less than the product of five and four? Write an expression.
(63)

3. Two thirds of the 12 guitar strings were out of tune. How many guitar strings were out of tune? Draw a picture to illustrate the problem.
(94)

*4. **Represent** Use digits to write eight million, nine hundred forty-five thousand in standard form. Then, round it to the nearest ten thousand.
(47, 48)

*5. **Connect** What is the sum of the numbers labeled A and B on the number line below?
(43, 45)

*6. Write a fraction equal to 1 and that has a denominator of 5.
(98)

7. **Represent** Use words to write $397\frac{3}{4}$.
(32)

8. Estimate the sum of 4178 and 6899 by rounding both numbers to the nearest thousand before adding.
(61)

*9. Change each improper fraction to a whole number or a mixed number:
(99)

a. $\frac{7}{3}$

b. $\frac{8}{4}$

c. $\frac{9}{5}$

*10. For the first 3 hours, the hikers hiked at 3 miles per hour. For the next
(58, 93) 2 hours they hiked at 4 miles per hour. If the total trip was 25 miles, how far did they still have to go?

*11. a. What fractional part of yard is a foot?
(42, 101)

    b. **(Formulate)** Write a formula for changing any number of yards to feet.

*12. **(Model)** Graph the relationship between yards and feet for 1, 2, and
(Inv. 10) 3 yards.

13. $41.6 + 13.17 + 9.2$
(45)

14. $h + 8.7 = 26.47$
(8, 45)

*15. $6\frac{3}{8} + 4\frac{2}{8}$
(100)

*16. $4\frac{7}{10} - 1\frac{6}{10}$
(100)

*17. We may write 48 as 40 + 8. Use the Distributive Property to find 5(40 + 8).
(102)

*18. **(Analyze)** Two fifths of the students rode the bus, and one fifth
(100) traveled by car. What fraction of the students either rode the bus or traveled by car?

19. $\$0.48 \times 5$
(38)

20. 80^2
(65, 85)

21. $\sqrt{25} \times \sqrt{25}$
(Inv. 3)

22. $4d = \$6.36$
(34, 79)

*23. $2\overline{)520}$
(79)

24. $\dfrac{175}{5}$
(69)

25. What is the perimeter and area of this square? Use
(66, 101) formulas to solve the problem.

10 in.

* 26. **Analyze** Use this coordinate graph to solve parts **a** and **b**.

(Inv. 9,
Inv. 10)

a. Which coordinates can you subtract to find the length of this segment? What is the length?

b. Name three points on this segment.

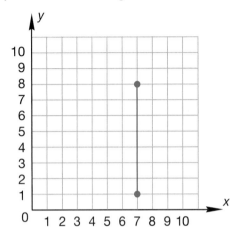

* 27. The tabletop was 76 cm above the floor. The tabletop was how many meters above the floor?

(42)

* 28. **Interpret** Use the line graph to answer parts **a–c.**

(Inv. 5,
Inv. 6)

**Average Summer Temperatures
in Huron, South Dakota**

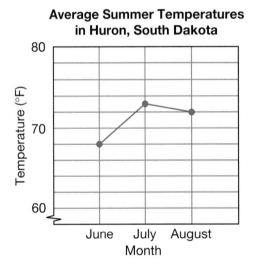

a. Write the names of the months in order from warmest to coolest.

b. How many degrees warmer is the average temperature during July than the average temperature during June?

c. Explain how to find the median temperature.

*29. There were $3\frac{4}{5}$ pitas in the chef's kitchen. Then the chef removed $1\frac{3}{5}$
(100) of the pitas. How many potpies remained in the chef's kitchen?

*30. Multiple Choice The mixed numbers $5\frac{3}{8}$ and $7\frac{4}{5}$ do not have common
(98) denominators, but we know their sum is between which two numbers?

  A 14 and 16   **B** 12 and 14
  C 10 and 12   **D** 5 and 8

Real-World Connection

Platonic solids, named after the Greek philosopher Plato, are a famous group of three-dimensional figures. The most commonly known Platonic solid is the cube (also called a hexahedron). The cube is, of course, based on a square and has 6 square faces. There are 3 Platonic solids based on a triangle, however, and 1 Platonic solid based on a pentagon. Refer to the table below.

Platonic Solid	Base Polygon	Faces	Edges	Vertices
Tetrahedron	Triangle	4	6	4
Cube	Square	6	12	8
Octahedron	Triangle	8	12	6
Dodecahedron	Pentagon	12	30	20
Icosahedron	Triangle	20	30	12

a. Which Platonic solid has the greatest number of vertices? Which has the fewest?

b. Name a major difference between an octagon and an octahedron.

c. Is an icosahedron always larger in size than a cube? Why?

d. What is another name for a tetrahedron? Draw a model of a tetrahedron.

California Mathematics Content Standards

NS 3.0, 3.2 Demonstrate an understanding of, and the ability to use, standard algorithms for multiplying a multidigit number by a two-digit number and for dividing a multidigit number by a one-digit number; use relationships between them to simplify computations and to check results.

MR 3.0, 3.2 Note the method of deriving the solution and demonstrate a conceptual understanding of the derivation by solving similar problems.

• Thinking About Multiplying Two-Digit Numbers

We can use a rectangle model to represent a multiplication problem. Let's look at 14 × 23.

We use expanded form to rewrite each of the factors.

$$14 = 10 + 4$$
$$23 = 20 + 3$$

Then we label the sides of a rectangle with the expanded form of the two factors. We multiply to find the area of each of the four rectangles that make up our larger rectangle.

	20	**+**	**3**
10	$\begin{array}{r} 20 \\ \times\ 10 \\ \hline 200 \end{array}$		$\begin{array}{r} 3 \\ \times\ 10 \\ \hline 30 \end{array}$
+ **4**	$\begin{array}{r} 20 \\ \times\ 4 \\ \hline 80 \end{array}$		$\begin{array}{r} 3 \\ \times\ 4 \\ \hline 12 \end{array}$

We add the areas
of the smaller
rectangles to find
the area of the
entire rectangle.

$$\begin{array}{r} 200 \\ 80 \\ 30 \\ +\ 12 \\ \hline 322 \end{array}$$

Explain Describe how you would draw a rectangle model for multiplying 24 × 48.

Apply Draw a rectangle model and use it to multiply 24 × 48.

California Mathematics Content Standards

NS 1.0, 1.5 Explain different interpretations of fractions, for example, parts of a whole, parts of a set, and division of whole numbers by whole numbers; explain equivalents of fractions (see Standard 4.0).

NS 1.0, 1.7 Write the fraction represented by a drawing of parts of a figure; represent a given fraction by using drawings; and relate a fraction to a simple decimal on a number line.

• Equivalent Fractions

facts   Power Up A

mental math

    a. Number Sense: $450 - 175$

    b. Number Sense: 50×42

    c. Money: Casius gave the clerk $2.00 for lemons that cost $1.62. How much change should he receive?

    d. Time: Which date occurs only once every four years?

    e. Powers/Roots: $2^3 \div 2$

    f. Estimation: Micalynn purchased 4 toothbrushes for $11.56. Round this amount to the nearest dollar and then divide by 4 to estimate the cost per toothbrush.

    g. Calculation: $\sqrt{36}$, $\times 3$, $+ 2$, $\div 10$, $- 1$

    h. Simplify: $4 - a$ when $a = 3$

problem solving   Choose an appropriate problem-solving strategy to solve this probelm. A square has 90° rotational symmetry because for every 90° it is turned, it matches its original position. A regular pentagon has rotational symmetry of 72°. Does a regular octagon have rotational symmetry? If so, what degree of rotational symmetry does it have?

New Concept

Math Language

Equivalent is another word for equal. For example, $\frac{1}{2}$ and $\frac{2}{4}$ are equivalent fractions, and $\frac{1}{2}$ and $\frac{2}{4}$ are equal fractions.

Equal portions of each circle below have been shaded. We see that different fractions are used to name the shaded portions.

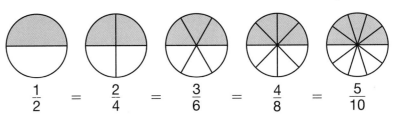

$$\frac{1}{2} = \frac{2}{4} = \frac{3}{6} = \frac{4}{8} = \frac{5}{10}$$

These fractions all name the same amount. Different fractions that name the same amount are called **equivalent fractions.**

Example 1

The rectangle on the left has three equal parts. We see that two parts are shaded, so two thirds of the figure is shaded.

$$\frac{2}{3} \qquad = \qquad \frac{?}{6}$$

The rectangle on the right has six equal parts. How many parts must be shaded so that the same fraction of this rectangle is shaded?

We see that **four parts** out of six must be shaded. This means two thirds is the same as four sixths.

$$\frac{2}{3} \qquad = \qquad \frac{4}{6}$$

$\frac{2}{3}$ and $\frac{4}{6}$ are equivalent fractions.

Example 2

What equivalent fractions are shown at right?

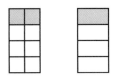

An equal portion of each rectangle is shaded. The rectangles shown are equal.

$$\frac{2}{8} = \frac{1}{4}$$

We remember that when we multiply a number by 1, the answer equals the number we multiplied.

$$2 \times 1 = 2 \qquad 2000 \times 1 = 2000 \qquad \frac{1}{2} \times 1 = \frac{1}{2}$$

We also remember that there are many ways to write "1."

$$1 = \frac{2}{2} = \frac{3}{3} = \frac{4}{4} = \frac{5}{5} = \frac{6}{6} = \cdots$$

We can use these two facts to find equivalent fractions. If we multiply a fraction by a fraction name for 1, the product is an equivalent fraction.

$$\frac{1}{2} \times \frac{2}{2} = \frac{2}{4} \qquad \begin{matrix}(1 \times 2 = 2) \\ (2 \times 2 = 4)\end{matrix}$$

By multiplying $\frac{1}{2}$ by $\frac{2}{2}$, which is a fraction name for 1, we find that $\frac{1}{2}$ equals $\frac{2}{4}$. Notice that we multiply numerator by numerator and denominator by denominator. We can find other fractions equal to $\frac{1}{2}$ by multiplying by other fraction names for 1:

$$\frac{1}{2} \times \frac{3}{3} = \frac{3}{6} \qquad \frac{1}{2} \times \frac{4}{4} = \frac{4}{8} \qquad \frac{1}{2} \times \frac{5}{5} = \frac{5}{10}$$

Example 3

Find four fractions equal to $\frac{1}{3}$ by multiplying $\frac{1}{3}$ by $\frac{2}{2}, \frac{3}{3}, \frac{4}{4},$ and $\frac{5}{5}$.

$$\frac{1}{3} \times \frac{2}{2} = \frac{2}{6} \qquad\qquad \frac{1}{3} \times \frac{3}{3} = \frac{3}{9}$$

$$\frac{1}{3} \times \frac{4}{4} = \frac{4}{12} \qquad\qquad \frac{1}{3} \times \frac{5}{5} = \frac{5}{15}$$

Each of our answers is a fraction equal to $\frac{1}{3}$.

Lesson Practice Name the equivalent fractions shown:

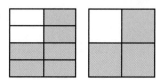

a.   b.

Draw pictures to show that the following pairs of fractions are equivalent:

c. $\frac{2}{4} = \frac{1}{2}$ **d.** $\frac{4}{6} = \frac{2}{3}$ **e.** $\frac{2}{8} = \frac{1}{4}$

Find four equivalent fractions for each fraction below. To do this, multiply each fraction by $\frac{2}{2}$, $\frac{3}{3}$, $\frac{4}{4}$, and $\frac{5}{5}$.

f. $\frac{1}{4}$ **g.** $\frac{5}{6}$ **h.** $\frac{2}{5}$ **i.** $\frac{1}{10}$

Written Practice

Distributed and Integrated

1. **Interpret** The pictograph shows the number of motor vehicles
(Inv. 5) that were driven past Cruz's home during 1 hour. Use the pictograph to answer the questions that follow.

Type of Vehicle	Number of Vehicles
Cars	⊙ ⊙ ⊙ ⊙ ⊙ ⊙
Trucks	⊙ ⊙
Mopeds	◖
Motorcycles	⊙ ◖

Key: ⊙ = 4 vehicles

a. What kind of vehicle was driven past Cruz's home two times?

b. Write a word sentence that compares the number of trucks to the number of cars.

c. Suppose ten bicyclists rode past Cruz's house. How many symbols would be needed to show the number of bicycles in the pictograph? Explain your answer.

*2. What number is six less than the sum of seven and eight? Write an
(63) expression.

*3. Nell read three tenths of 180 pages in one day. How many pages did
(94) she read in one day?

4. a. The thermometer shows the temperature of a warm
(21)   October day in Buffalo, New York. What temperature
        does the thermometer show?

b. On a cold day in January an afternoon temperature
    could be 19°F. If the temperature dropped 22 degrees
    overnight, what would the temperature be?

5. A circular disk, divided into 8 equal pieces, represents what fraction
(98) name for 1?

6. a. What is the diameter of this dime?
(18, 42)

b. What is the radius of the dime?

c. What is the diameter of the dime in centimeters?

7. There are 11 players on a football team, so when two teams play, there
(37, 38, are 22 players on the field at one time. Across the county on a Friday
 39) night in October, many games are played. The table shows the number
of players on the field for a given number of games. How many players
are on the field in 5 games? 10 games?

Number of games	1	2	3	4	5
Number of players	22	44	66	88	?

8. Rick left home at the time shown on the clock and arrived
(13) at a friend's house 15 minutes later. At what time did Rick
arrive at his friend's house?

*9. (**Represent**) Change the improper fraction $\frac{5}{2}$ to a mixed number. Draw
(99) a picture that shows that the improper fraction and the mixed number
are equal.

*10. Use the information below to answer parts **a** and **b**.

(11, 76, Inv. 10)

> Chico did 12 push-ups on the first day. On each of the next four days, he did two more push-ups than he did the day before.

  a. Altogether, Chico did how many push-ups in five days?

  b. Make a table to represent the problem.

*11. (**Interpret**) Name two points on this line.

(Inv. 9, Inv. 10)

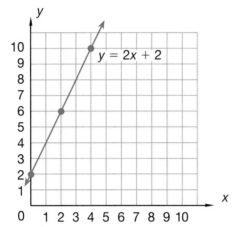

*12. (**Analyze**) There are blue, orange, and grey marbles in a bag. There are twice as many orange marbles as there are blue marbles. There are two less grey marbles than orange. If there are 8 grey marbles, then how many blue are there?

(64, 93)

*13. Find three fractions equivalent to $\frac{2}{3}$ by multiplying $\frac{2}{3}$ by $\frac{2}{2}$, $\frac{3}{3}$, and $\frac{10}{10}$.

(103)

*14. Since 63 equals 60 + 3, we may find 5 × 63 by finding 5(60 + 3). Use the Distributive Property to find 5(60 + 3).

(102)

*15. Find *ac* when *a* is 18 and *c* is 22.

(63, 88)

*16. (**Formulate**) A store owner had expenses of $12,000 and sales of $8000. Write a formula to show the profit or loss.

(101)

17. Find the median, mode, and range of this set of scores:

(Inv. 6, 95)

$$100,\ 100,\ 95,\ 90,\ 90,\ 80,\ 80,\ 80,\ 60$$

*18. **Multiple Choice** If a quadrilateral has two pairs of parallel sides, then the quadrilateral is certain to be a _____.

(90)

  A rectangle   **B** parallelogram

  C trapezoid   **D** square

19. $v + 8.5 = 24.34$
(8, 45)

20. $26.4 - 15.18$
(45)

21. $4 \times 3 \times 2 \times 1$
(23)

22. 26×30
(71)

23. $8)\overline{\$16.48}$
(79, 83)

*24. $10n = 250$
(34, 85)

*25. $\dfrac{5}{12} + \dfrac{6}{12}$
(100)

*26. $\dfrac{8}{12} - \dfrac{3}{12}$
(100)

27. How many square feet of paper are needed to cover a bulletin board
(Inv. 3) that is 3 feet tall and 6 feet wide?

*28. The bread recipe calls for $7\frac{1}{2}$ cups of flour to make 2 loaves of bread.
(93, 100) The baker wants to make 4 loaves of bread. How many cups of flour does the baker need?

*29. The backpackers camped in a tent. Refer to the figure at right
(96) to answer parts **a–c**.

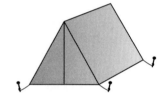

  a. The tent has the shape of what geometric solid?

  b. Including the bottom, how many faces does it have?

  c. How many edges does it have?

30. The flag of the United States has thirteen stripes. Six of the stripes are
(19, 76) white, and the rest of the stripes are red.

  a. How many red stripes are on the American flag?

  b. What fraction of the stripes on the American flag are white?

  c. What fraction of the stripes on the American flag are red?

California Mathematics Content Standards

NS 1.0, 1.5 Explain different interpretations of fractions, for example, parts of a whole, parts of a set, and division of whole numbers by whole numbers; explain equivalents of fractions (see Standard 4.0).

MR 1.0, 1.1 Analyze problems by identifying relationships, distinguishing relevant from irrelevant information, sequencing and prioritizing information, and observing patterns.

• Use Equivalent Fractions to Find Common Denominators

Equivalent fractions are fractions that represent the same number.

When we add or subtract fractions, we write the fractions with common denominators first. There are a few ways to do this.

Method 1 Multiply the denominators.

$\frac{2}{3} + \frac{1}{5}$	We want the fractions to have the same denominator.
$3 \times 5 = 15$	We multiply the denominators to find the common denominator.
$\frac{2 \times 5}{3 \times 5}$ and $\frac{1 \times 3}{5 \times 3}$	We must multiply the numerator and denominator by the same number.
$\frac{10}{15} + \frac{3}{15} = \frac{13}{15}$	The two fractions now have a common denominator of 15. Now we add.

Method 2 Find the least common multiple of each denominator.

$\frac{1}{6} + \frac{3}{4}$	We want the fractions to have the same denominator.
$\frac{1}{6} \cdot \frac{2}{2} + \frac{3}{4} \cdot \frac{3}{3}$	12 is a multiple of 6 and 4. We can use 12 for the common denominator instead of 24.
$\frac{2}{12} + \frac{9}{12} = \frac{11}{12}$	The two renamed fractions have the same denominator. We can add.

Apply Decide the best way to find a common denominator.

a. $\frac{3}{4} - \frac{5}{12}$

b. $\frac{3}{5} + \frac{1}{3}$

California Mathematics Content Standards

NS 1.0, **1.1** Read and write whole numbers in the millions.

NS 1.0, **1.3** Round whole numbers through the millions to the nearest ten, hundred, thousand, ten thousand, or hundred thousand.

MR 2.0, 2.3 Use a variety of methods, such as words, numbers, symbols, charts, graphs, tables, diagrams, and models, to explain mathematical reasoning.

• Rounding Whole Numbers Through Hundred Millions

Power Up

facts　　Power Up A

mental math　　Think of one cent more or less than quarters in problems **a–c.**

a. Number Sense: 425 + 374

b. Number Sense: 550 − 324

c. Number Sense: $4.49 + $2.26

d. Number Sense: 15 × 40

e. Time: Each section of the test takes 25 minutes. There is a 5-minute break between sections. If the class starts the test at 9:00 a.m., how many sections can the class finish by 10:30 a.m.?

f. Estimation: Estimate 35 × 25. Round 35 to 40, round 25 down to 20, and then multiply.

g. Calculation: 2 × 2, square the number, + 4, ÷ 5, − 4

h. Simplify: $2b$ when $b = 5$

problem solving　　Choose an appropriate problem-solving strategy to solve this problem. Todd rode his bicycle down his 50-foot driveway and counted eight full turns of the front wheel. How many times will the front wheel turn if he rides 100 yards?

New Concept

In this lesson we will practice rounding large numbers to the nearest ten thousand, the nearest hundred thousand, and so on through the nearest hundred million.

Recall the locations of the whole-number place values through hundred millions:

Whole Number Place Values

| hundred trillions | ten trillions | trillions | | hundred billions | ten billions | billions | | hundred millions | ten millions | millions | | hundred thousands | ten thousands | thousands | | hundreds | tens | ones | decimal point |

___ ___ ___ , ___ ___ ___ , ___ ___ ___ , ___ ___ ___ , ___ ___ ___ .

After rounding to the nearest ten thousand, each place to the right of the ten-thousands place will be zero.

Analyze How is the value of each place related to the value of the place to its right?

Example 1

Round 12,876,250 to the nearest million.

The number begins with "twelve million." Counting by millions from 12 million, we say "twelve million, thirteen million," and so on. We know that 12,876,250 is between 12 million and 13 million. Since 12,876,250 is more than halfway to 13 million, we round up to **13,000,000.**

Example 2

Round 16,458,500 to the nearest hundred thousand.

Since we want to round to the nearest hundred thousand, we look at the place to the right of the hundred thousands place—the ten thousands place.

ten thousands place

16, 4⑤8,500

hundred thousands place

The digit in the ten thousands place is 5 so we add 1 to the digit in the hundred thousands place. All the digits to the right of the hundred thousands place become zeros.

To the nearest hundred thousand, 16,458,500 rounds to **16,500,000.**

Example 3

Round 237,984,000 to the nearest hundred million.

Since we want to round to the nearest hundred million, we look at the place to the right of the hundred millions place—the ten millions place.

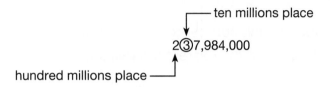

ten millions place

2③7,984,000

hundred millions place

The digit in the ten millions place is 3 so we leave the digit in the hundred millions place unchanged. All the digits to the right of the hundred millions place become zeros.

To the nearest hundred million, 237,984,000 rounds to **200,000,000.**

Lesson Practice

a. Round 2,156,324 to the nearest million.

b. Round 28,376,000 to the nearest ten million.

c. Round 412,500,000 to the nearest hundred million.

Estimate Round each number to the nearest hundred thousand:

d. 5,346,891   **e.** 75,965,000   **f.** 350,525,000

Written Practice   *Distributed and Integrated*

* **1.** **Analyze** Eighty students were divided among three classrooms as
(87)   equally as possible. Write three numbers to show how many students
were in each of the three classrooms.

* **2.** **Formulate** When the sum of three and four is subtracted from the
(63)   product of three and four, what is the difference? Write an equation to
solve the problem.

3. **Explain** Irma is twice as old as her sister and three years younger
(93)   than her brother. Irma's sister is six years old. How old is Irma's
brother? What is the first step?

* **4.** Four ninths of 513 fans cheered when the touchdown was scored. How
(94)   many fans cheered?

*5. **Connect** These circles show fractions equivalent to $\frac{1}{2}$.
(103)   Name the fractions shown.

*6. **Analyze** Use the Distributive Property to multiply 34 and 65.
(102)

*7. **Represent** Round each number to the given place:
(104)
   a. Round 13,458,912 to the nearest ten million.

   b. Round 13,458,912 to the nearest million.

   c. Round 13,458,912 to the nearest hundred thousand.

   d. Round 13,458,912 to the nearest ten thousand.

*8. **Explain** In a sporting goods store, an aluminum baseball bat sells
(61)   for $38.49, a baseball sells for $4.99, and a baseball glove sells for
   $24.95. What is a reasonable estimate of the cost of a bat, a glove,
   and two baseballs? Explain why your estimate is reasonable.

*9. Change the improper fraction $\frac{5}{2}$ to a mixed number.
(99)

10. Paul ran 7 miles in 42 minutes. What was the average number of minutes
(60)   it took Paul to run one mile?

11. Kia bought 3 scarves priced at $2.75 each. Tax was 58 cents. How much
(101)   did she pay for the scarves? Write a formula to solve the problem.

12. **Analyze** Two tickets for the play cost $26. At that rate, how much
(93)   would twenty tickets cost?

*13. Dawn is $49\frac{1}{2}$ inches tall. Tim is $47\frac{1}{2}$ inches tall. Dawn is how many
(100)   inches taller than Tim?

14. 7.43 + 6.25 + 12.7
(45)

15. $q + 7.5 = 14.36$
(8, 45)

16. 90 × 800
(85)

17. $f \times 73$¢ if $f = 8$
(28, 38)

18. 7 × 6 × 5 × 0
(23)

19. 15^2
(65, 88)

20. 60×5^2
(65, 71)

21. $\sqrt{49} \times \sqrt{49}$
(Inv. 3)

*22. $5\frac{1}{3} + 3\frac{1}{3}$
(100)

*23. $4\frac{4}{5} - 3\frac{3}{5}$
(100)

*24. $\dfrac{1242}{6}$
(83)

*25. $4\overline{)3000}$
(83)

26. This square has a perimeter of 8 cm. Find the length of each
(20, 66) side. Then find the area of the square.

*27. Refer to this bus schedule to answer parts **a–c.**
(RF13, 13)

Route 346

Destination	Arrival Time	Arrival Time	Arrival Time
	6:43 a.m.	7:25 a.m.	3:45 p.m.
5th & Western	6:50 a.m.	7:32 a.m.	3:50 p.m.
5th & Cypress	6:54 a.m.	7:36 a.m.	3:55 p.m.
Cypress & Hill	7:01 a.m.	7:43 a.m.	4:03 p.m.
Hill & Lincoln	7:08 a.m.	7:50 a.m.	4:12 p.m.
Lincoln & 5th	7:16 a.m.	7:58 a.m.	4:20 p.m.

a. Ella catches the 6:50 a.m. bus at 5th and Western. When can she expect to arrive at Hill and Lincoln?

b. If the bus runs on schedule, how many minutes is her ride?

c. If Ella misses the 6:50 a.m. bus, then when can she catch the next Route 346 bus at that corner?

28. When Xena says a number, Yoli doubles the number and adds 3. Xena
(64, 93) and Yoli record their numbers in a table.

X	1	2	5	7
Y	5	7	13	17

a. (Predict) What number does Yoli say if Xena says 11?

b. (Generalize) Write an equation that shows the relationship of X and Y.

* 29. (**Interpret**) Use the graph below to solve parts **a–b.**
(Inv. 10)

    a. Kyle earns $5 per hour packing boxes. How long will it take him to earn $50?

    b. Write an equation to show the relationship between hours and dollars.

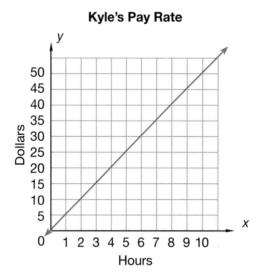

Kyle's Pay Rate

* 30. (**Represent**) A variety of morning times and temperatures are shown
(Inv. 5) in the table below:

Morning Temperatures

Time	Temperature (°F)
12:00 a.m.	51
2:00 a.m.	48
4:00 a.m.	49
6:00 a.m.	50
8:00 a.m.	56
10:00 a.m.	62

Display the data in a line graph. Then write one statement that describes the data.

LESSON
105

🖊 *California Mathematics Content Standards*

NS 4.0, 4.1 Understand that many whole numbers break down in different ways (e.g., $12 = 4 \times 3 = 2 \times 6 = 2 \times 2 \times 3$).

NS 4.0, 4.2 Know that numbers such as 2, 3, 5, 7, and 11 do not have any factors except 1 and themselves and that such numbers are called prime numbers.

• Factoring Whole Numbers

facts	Power Up A
mental math	Think of one cent more or less than quarters in problems **a–c**.

    a. Number Sense: $126 + 375$

    b. Number Sense: $651 - 225$

    c. Number Sense: $\$6.51 + \$2.75$

    d. Money: The atlas cost $16.25. Amol paid for it with a $20 bill. How much change should he receive?

    e. Measurement: Fran drank $1\frac{1}{2}$ quarts of water. How many pints did she drink?

    f. Estimation: Estimate 32×28.

    g. Calculation: $40 \div 4, \times 6, + 4, \sqrt{\ }, - 8$

    h. Simplify: $d + 6$ when $d = 8$

problem solving

Choose an appropriate problem-solving strategy to solve this problem. This sequence has an alternating pattern. Copy this sequence on your paper, and continue the sequence to 18. Then describe the pattern in words.

$$0, 5, 3, 8, 6, 11, 9, 14, \ldots$$

New Concept

Math Language

Recall that a prime number is a number that has only two factors itself and 1.

We have learned to write a number as the product to two factors.

$$24 = 1 \times 24, 2 \times 12, 3 \times 8, \text{ and } 4 \times 6$$

Today, we will learn how to write a number as the product of three or more factors.

Below are some ways we can write 24 as the product of three factors.

$$24 = 2 \times 2 \times 6, \ 1 \times 12 \times 2, \text{ and } 2 \times 3 \times 4$$

Another way to write 24 is as the product of *prime factors.*

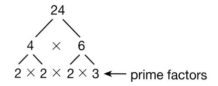

$$2 \times 2 \times 2 \times 3 \longleftarrow \text{prime factors}$$

Example 1

Write 40 as the product of two factors in as many different ways as you can. Then list all the factors of 40.

We know that 1 is a factor of every number, so, we start with 1 and 40.

Since 40 is an even number, we know that 2 is a factor, $2 \times 20 = 40$.

Since 40 ends with a zero, we know that 5 and 10 are factors.

$$5 \times 8 = 40 \qquad 4 \times 10 = 40$$

We check and find that 3, 6, 7, and 9 are not factors of 40.

The factor pairs of 40 are, **1 × 40, 2 × 20, 4 × 10, and 5 × 8.**
The factors of 40 are, **1, 2, 4, 5, 8, 10, 20, and 40.**

Example 2

Write 40 as the product of prime factors.

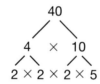

$$2 \times 2 \times 2 \times 5$$

We start by finding two factors for 40 such as, 2 and 20.

Since 2 is a prime number, find two factors for 20 such as, 2 and 10.

Again, 2 is a prime number, find two factors for 10 such as, 2 and 5. Both 2 and 5 are prime numbers.

The number 40 is a product of the prime factors:

$$2 \times 2 \times 2 \times 5$$

Lesson Practice   **Analyze** Write each number as a product of two numbers in as many different ways as you can:

  a. 16   **b.** 20

  c. **Verify** Write 27 as the product of prime factors.

Written Practice   *Distributed and Integrated*

---

*1. a. Five minutes is equal to how many seconds? (*Hint:* There are 60
(13, 71, 85)   seconds in each minute.)

  b. Sixty minutes is how many seconds?

*2. **Explain** Trevor, Ann, and Lee were playing marbles. Ann had twice
(93)   as many marbles as Trevor had, and Lee had 5 more marbles than Ann had. Trevor had 9 marbles. How many marbles did Lee have? What is the first step?

3. On each of 5 bookshelves there are 44 books. How many books are on
(39)   all 5 bookshelves?

*4. a. Nine tenths of the 30 students remembered their homework. How
(67, 94)   many students remembered their homework?

  b. What fractional part of the students did not remember their homework?

5. For parts **a–c,** refer to this number line:
(43)

  a. The number for point *A* is what fraction?

  b. The number for point *B* is what decimal number?

  c. The number for point *C* is what fraction?

6. What fraction name for 1 has a denominator of 3?
(98)

*7. What equivalent fractions are shown below?
(103)

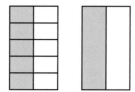

*8. **Represent** Draw a picture to show that $\frac{6}{8}$ and $\frac{3}{4}$ are equivalent
(Inv. 8
103) fractions.

9. Below is a golf scorecard for 9 holes of miniature golf:
(Inv.6,
95)

Putt 'N' Putt

Player	1	2	3	4	5	6	7	8	9	Total
Michelle	6	7	5	2	4	1	3	5	3	36
Mathea	5	4	4	3	4	3	2	5	3	33

What was Michelle's median score? What was the mode?

10. It was 11:00 a.m., and Sarah had to clean the laboratory by 4:20 p.m.
(13) How much time did she have to clean the lab?

*11. Draw a quadrilateral that has two sides that are parallel, a third side
(17, 50) that is perpendicular to the parallel sides, and a fourth side that is
not perpendicular to the parallel sides. What type of quadrilateral did
you draw?

12. The factors of 10 are 1, 2, 5, 10. The factors of 15 are 1, 3, 5, 15. Which
(105) number is the largest factor of both 10 and 15?

13. **List** What are the prime factors of 8.
(105)

14. $4.3 + 12.6 + 3.75$ **15.** $364.1 - 16.41$
(45)   (45)

*16. $\frac{5}{8} + \frac{2}{8}$ ***17.** $\frac{3}{5} + \frac{1}{5}$ ***18.** $1\frac{9}{10} - 1\frac{2}{10}$
(100)   (100)   (100)

19. 60×800 **20.** 73×48 **21.** $9 \times 78¢$
(85)   (88)   (28, 38)

22. 10^3 **23.** $4x = 3500$ **24.** $\frac{4824}{8}$
(65, 85)   (34, 79)   (83)

*25. $6\overline{)540}$ ***26.** $8\overline{)463}$
(75)   (72)

*27. Estimate the perimeter and area of this figure. Each small square
_(RF14) represents one square inch.

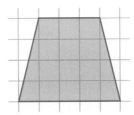

*28. (**Represent**) Draw a rectangle that is 4 cm long and 1 cm wide. Then
_(18, Inv. 4) shade 0.25 of it.

29. Multiple Choice Which of the following is a cylinder?
₍₉₆₎

A 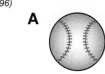　　B 　　C 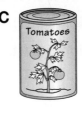　　D

*30. (**Analyze**) Hans divided 356 by 4 and checked his answer using
₍₁₀₂₎ the Distributive Property. Show how to check his answer with the
Distributive Property.

California Mathematics Content Standards

NS 1.0, 1.7 Write the fraction represented by a drawing of parts of a figure; represent a given fraction by using drawings; and relate a fraction to a simple decimal on a number line.

NS 4.0, 4.2 Know that numbers such as 2, 3, 5, 7, and 11 do not have any factors except 1 and themselves and that such numbers are called prime numbers.

MR 2.0, 2.3 Use a variety of methods, such as words, numbers, symbols, charts, graphs, tables, diagrams, and models, to explain mathematical reasoning.

• Reducing Fractions

facts   Power Up A

mental math Find each fraction of 24 in problems **a–c.**

  a. Fractional Part: $\frac{1}{2}$ of 24

  b. Fractional Part: $\frac{1}{3}$ of 24

  c. Fractional Part: $\frac{1}{4}$ of 24

  d. Number Sense: 4×18

  e. Money: Stefano has $3.75 in his pocket and $4.51 in his piggy bank. Altogether, how much money does Stefano have?

  f. Estimation: Estimate 62×19.

  g. Calculation: 5^2, $+ 10$, $- 3$, $\div 4$, $\times 2$

  h. Simplify: $e \div 3$ when $e = 30$

problem solving Choose an appropriate problem-solving strategy to solve this problem. Two cups make a pint. Two pints make a quart. Two quarts make a half gallon, and two half gallons make a gallon. A pint of water weighs about one pound. Find the approximate weight of a cup, a quart, a half gallon, and a gallon of water.

New Concept

Recall from Investigation 8 that when we *reduce* a fraction, we find an equivalent fraction written with smaller numbers or **lowest terms**. The picture below shows $\frac{4}{6}$ reduced to $\frac{2}{3}$.

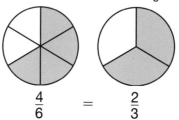

$$\frac{4}{6} \quad = \quad \frac{2}{3}$$

Not all fractions can be reduced. Only a fraction whose numerator and denominator can be divided by the same number can be reduced. Since both the numerator and denominator of $\frac{4}{6}$ can be divided by 2, we can reduce the fraction $\frac{4}{6}$.

To reduce a fraction, we will use a fraction that is equal to 1. To reduce $\frac{4}{6}$, we will use the fraction $\frac{2}{2}$. We divide both 4 and 6 by 2, as shown.

$$\frac{4}{6} = \frac{4 \div 2}{6 \div 2} = \frac{2}{3}$$

Example

Thinking Skills

Discuss

How do we know that both 6 and 8 are divisible by 2?

Write the reduced form of each fraction:

a. $\frac{6}{8}$ **b.** $\frac{3}{6}$ **c.** $\frac{6}{7}$

a. The numerator and denominator are 6 and 8. These numbers can be divided by 2. That means we can reduce the fraction by dividing 6 and 8 by 2.

$$\frac{6}{8} = \frac{6 \div 2}{8 \div 2} = \frac{3}{4}$$

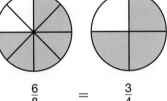

$$\frac{6}{8} \quad = \quad \frac{3}{4}$$

b. The numerator and denominator are 3 and 6. These numbers can be divided by 3, so we reduce $\frac{3}{6}$ by dividing both 3 and 6 by 3.

$$\frac{3}{6} = \frac{3 \div 3}{6 \div 3} = \frac{1}{2}$$

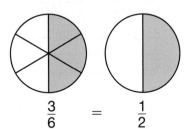

$$\frac{3}{6} \quad = \quad \frac{1}{2}$$

c. The numerator is 6 and the denominator is 7. The only number that divides 6 and 7 is 1. Dividing the terms of a fraction by 1 does not reduce the fraction.

$$\frac{6}{7} = \frac{6 \div 1}{7 \div 1} = \frac{6}{7}$$

The fraction $\frac{6}{7}$ cannot be reduced.

Justify Which number, 6 or 7, is prime? Explain why.

Lesson Practice Write the reduced form of each fraction:

a. $\frac{2}{4}$ **b.** $\frac{2}{6}$ **c.** $\frac{3}{9}$ **d.** $\frac{3}{8}$

e. $\frac{2}{10}$ **f.** $\frac{4}{10}$ **g.** $\frac{9}{12}$ **h.** $\frac{9}{10}$

i. $\frac{2}{8}$ **j.** $\frac{5}{10}$ **k.** $\frac{10}{12}$ **l.** $\frac{6}{10}$

Written Practice   *Distributed and Integrated*

*1. Use the following information to answer parts **a** and **b**.
(76, 93)

   One fence board costs 90¢. It takes 10 boards to build 5 feet of fence.

   a. How many boards are needed to build 50 feet of fence?

   b. How much will the boards cost altogether?

2. Use a formula to find the perimeter and area of this
(66) rectangle.

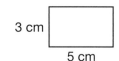

3 cm

5 cm

3. a. Find the length of this line segment in millimeters.
(42)

   b. Find the length of the segment in centimeters.

mm 10   20   30   40   50

cm   1   2   3   4   5

*4. Five ninths of the 36 horses were gray. How many of the horses
(94) were gray?

*5. Change each improper fraction to a whole number or a mixed number:
(99)
    a. $\frac{15}{2}$ **b.** $\frac{15}{3}$ **c.** $\frac{15}{4}$

*6. Angelina's mom is more than 32 years old but less than 40 years
(56) old, and her age in years is a prime number. How old is Angelina's
mom?

*7. What equivalent fractions are shown at right?
(Inv. 8,
103)

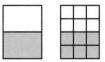

*8. A regular polygon has all sides the same length and all angles the
(50, 82, same measure.
90)

    a. Draw a regular quadrilateral. Show all the lines of symmetry.

    b. A regular quadrilateral has how many lines of symmetry?

    c. Does a regular quadrilateral have rotational symmetry?

*9. Write the reduced form of each fraction:
(106)
    a. $\frac{3}{6}$ **b.** $\frac{4}{6}$ **c.** $\frac{6}{12}$

*10. (List) What are the prime factors of 30?
(105)

11. T-shirts were priced at $5 each. Yoshi had $27 and bought 5 T-shirts.
(93) Tax was $1.50. How much money did he have left?

*12. $3\frac{3}{9} + 4\frac{4}{9}$ ***13.** $\frac{1}{7} + \frac{2}{7} + \frac{3}{7}$ **14.**   37.2
(100)   (100)   (45)   135.7
                                                           10.62

*15. $\frac{11}{12} - \frac{10}{12}$ ***16.** $\frac{8}{10} - \frac{5}{10}$ 2.47
(100)   (100)   + 14.0

17.   48   **18.**   72   **19.** $4.08
(88) $\times$ 36   (88) $\times$ 58   (59) $\times$ 7

20. 25.42 + 24.8
(45)

21. 36.2 − 4.27
(45)

*22. 90 ÷ 2
(68)

23. $\frac{5}{8} - \frac{5}{8}$
(100)

24. 7)2549
(79)

*25. $19.40 ÷ 4
(79)

*26. Write the reduced form of each fraction.
(106)

    a. $\frac{6}{9}$ **b.** $\frac{2}{12}$ **c.** $\frac{10}{15}$

*27. a. What is the geometric name for the shape of this box?
(96, 97)

    b. True or false: All of the opposite faces of the box are parallel.

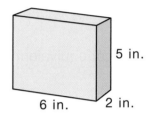

5 in.

6 in.   2 in.

*28. (**Represent**) Round each number to the given place:
(104)

    a. Round 241,679,500 to the nearest hundred million.

    b. Round 241,679,500 to the nearest ten million.

    c. Round 241,679,500 to the nearest million.

    d. Round 241,679,500 to the nearest hundred thousand.

*29. (**Evaluate**) If $y = 2x + 8$, what is y when x is 4, 5, and 6?
(64)

*30. (**Explain**) Estimate the perimeter and area of this shoe print. Each small square represents one square inch. Describe the method you used.
(RF14)

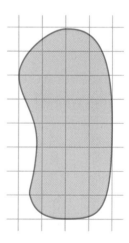

✎ **California Mathematics Content Standards**

NS **3.0, 3.3** Solve problems involving multiplication of multidigit numbers by two-digit numbers.

MR 2.0, 2.3 Use a variety of methods, such as words, numbers, symbols, charts, graphs, tables, diagrams, and models, to explain mathematical reasoning.

• Multiplying a Three-Digit Number by a Two-Digit Number

facts   Power Up B

mental math   Find each fraction of 30 in problems **a–c.**

    a. Fractional Part: $\frac{1}{2}$ of 30

    b. Fractional Part: $\frac{1}{3}$ of 30

    c. Fractional Part: $\frac{1}{5}$ of 30

    d. Number Sense: 50×28

    e. Time: The soccer match ended at 1:15 p.m. The match had started $1\frac{1}{2}$ hours earlier. When did the match begin?

    f. Estimation: To estimate 26×19, round 26 down to 25, round 19 up to 20, and then multiply.

    g. Calculation: $5 \times 2, \times 10, \div 2, -1, \sqrt{}$

    h. Simplify: s^2 when $s = 3$

problem solving   Choose an appropriate problem-solving strategy to solve this problem. In parts of the country where "daylight savings time" is observed, we follow the rule "spring forward, fall back." This rule means we turn the clock forward one hour in the spring and back one hour in the fall. Officially, clocks are reset at 2 a.m. on a Sunday. How many hours long are each of those Sundays when the clocks are reset?

New Concept

We have learned to multiply a two-digit number by another two-digit number. In this lesson we will learn to multiply a three-digit number by a two-digit number.

Example 1

A bakery is open 364 days each year. On each of those days, the bakery owner bakes 24 loaves of bread. How many loaves of bread does the owner bake each year?

Thinking Skills

Justify

Why are there two partial products?

We write the three-digit number above the two-digit number so that the last digits in each number are lined up. We multiply 364 by 4. Next we multiply 364 by 2. Since this 2 is actually 20, we write the last digit of this product in the tens place, which is under the 2 in 24. Then we add and find that the owner bakes **8736 loaves of bread** each year.

$$
\begin{array}{r}
1 \\
2\,1 \\
364 \\
\times\ \ 24 \\
\hline
1456 \\
728\ \ \\
\hline
8736
\end{array}
$$

Example 2

During summer vacation, a school principal ordered 38 paperback dictionaries for the school bookstore. The cost of each dictionary was $4.29. What was the total cost of the dictionaries?

Thinking Skills

Generalize

When one factor of a multiplication is dollars and cents, how many decimal places will be in the product? Name the places.

We will ignore the dollar sign and decimal point until we are finished multiplying. First we multiply 429 by 8. Then we multiply 429 by 3 (which is actually 30), remembering to shift the digits of the product one place to the left. We add and find that the product is 16302. Now we write the dollar sign and insert the decimal point two places from the right. We find that the total cost of the dictionaries was **$163.02.**

$$
\begin{array}{r}
2 \\
2\ 7 \\
\$4.29 \\
\times\ \ \ 38 \\
\hline
34.32 \\
128.7\ \ \\
\hline
\$163.02
\end{array}
$$

Lesson Practice   Multiply:

a. 235×24 **b.** 14×430 **c.** $\$1.25 \times 24$

d. $\begin{array}{r} 416 \\ \times\ 32 \\ \hline \end{array}$ **e.** $\begin{array}{r} \$6.25 \\ \times\ \ \ 31 \\ \hline \end{array}$ **f.** $\begin{array}{r} 562 \\ \times\ 47 \\ \hline \end{array}$

1. Carrie drove to visit her cousin, who lives 3000 miles away. If Carrie
(51, 52, 93) drove 638 miles the first day, 456 miles the second day, and 589 miles the third day, how much farther does she need to drive to get to her cousin's house?

2. Use a formula to find the perimeter and area of this
(66, 101) square:

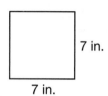

7 in.

7 in.

3. If the perimeter of a square is 2 meters, then each side is how many
(20, 42) centimeters long?

*4. **Represent** Round 917,250,000 to the nearest hundred million.
(104)

5. Round 6843 to the nearest thousand.
(40)

*6. Write the reduced form of each fraction:
(106)
    a. $\frac{4}{5}$ **b.** $\frac{5}{10}$ **c.** $\frac{4}{10}$

7. **Represent** Write 374.25 using words.
(Inv. 4)

*8. **Represent** Draw a picture to show that $\frac{1}{2}$ and $\frac{4}{8}$ are equivalent
(19, Inv. 8, 103) fractions.

*9. **Connect** Write three fractions equivalent to $\frac{1}{4}$.
(103)

10. **Estimate** The concession stand at an elementary school basketball
(91) tournament earned a profit of $850 during a 3-day tournament. About how much profit was earned each day?

*11. **Analyze** The explorer Zebulon Pike estimated that the mountain's
(15, 51) height was eight thousand, seven hundred forty-two feet. His estimate was five thousand, three hundred sixty-eight feet less than the actual height. Today we call this mountain Pikes Peak. What is the height of Pikes Peak?

12. $6\overline{)4837}$
(83)

13. $\dfrac{1372}{\sqrt{16}}$
(Inv. 3, 79)

*14. $4\overline{)960}$
(83)

*15. $5\overline{)1360}$
(79)

16. $30.07 - 3.7$
(45)

17. $46.0 - 12.46$
(45)

18.
(45)

$$\begin{array}{r} 37.15 \\ 6.84 \\ 1.29 \\ 29.1 \\ +\ 3.6 \\ \hline \end{array}$$

*19.
(107)

$$\begin{array}{r} \$3.28 \\ \times\quad 46 \\ \hline \end{array}$$

*20.
(107)

$$\begin{array}{r} 345 \\ \times\quad 25 \\ \hline \end{array}$$

*21. $\dfrac{8}{15} + \dfrac{6}{15}$
(100)

*22. $4\dfrac{4}{5} - 1\dfrac{3}{5}$
(100)

*23. Estimate the perimeter and area of this triangle. Each
(RF14) small square represents one square centimeter.

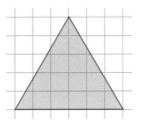

24. (**Conclude**) Write the next three numbers in this counting sequence:
(1, 27)
$$\ldots, 10{,}000, 20{,}000, 30{,}000, \ldots$$

*25. a. Multiple Choice Which of these triangles appears to be an
(17, 18, 81) equilateral triangle?

A   **B**   **C**   **D**

b. Describe the angles in triangle **B.**

c. Describe the segments in triangle **B.**

26. Multiple Choice To remove the lid from the pickle jar, Nadir turned
(78) the lid counterclockwise two full turns. Nadir turned the lid about how many degrees?

A 360°   **B** 180°   **C** 720°   **D** 90°

27. a. Which of these letters has no lines of symmetry?
(82)

M I C K E Y

b. Which letter has rotational symmetry?

* **28.** If each side of an equilateral triangle is $2\frac{1}{4}$ inches long, what is the
(20,
100) perimeter of the triangle?

* **29.** (List) **a.** Write 35 as the product of two factors as many ways as
(55, 56,
105) possible.

   b. Write 35 as the product of prime factors.

* **30.** (Verify) Use the Distributive Property to multiply 23 and 36. Show
(102) your work.

LESSON

108

California Mathematics Content Standards

MG 3.0, 3.1 Identify lines that are parallel and perpendicular.

MG 3.0, 3.3 Identify congruent figures.

MG 3.0, 3.6 Visualize, describe, and make models of geometric solids (e.g., prisms, pyramids) in terms of the number and shape of faces, edges, and vertices; interpret two-dimensional representations of three-dimensional objects; and draw patterns (of faces) for a solid that, when cut and folded, will make a model of the solid.

• Analyzing Prisms

facts	Power Up B
mental math	Find each fraction of 36 in problems **a–c**.

    a. Fractional Part: $\frac{1}{2}$ of 36

    b. Fractional Part: $\frac{1}{3}$ of 36

    c. Fractional Part: $\frac{1}{4}$ of 36

    d. Number Sense: $83 - 68$

    e. Geometry: What is the perimeter of a hexagon with sides that are each 5 cm long?

    f. Estimation: Camille is cutting lengths of yarn that are each $7\frac{3}{4}$ inches long. If she must cut 6 pieces of yarn, about how many inches of yarn will she need?

    g. Calculation: $10 \div 2$, $\times 8$, $- 4$, $\div 6$

    h. Simplify: c^2 when $c = 4$

problem solving

Choose an appropriate problem-solving strategy to solve this problem. In this sequence, each term is the sum of the two preceding terms. Copy this sequence and find the next four terms.

$$1, 1, 2, 3, 5, 8, \underline{\quad}, \underline{\quad}, \underline{\quad}, \underline{\quad}, \ldots$$

New Concept

A **prism** is a three-dimensional solid with two congruent bases. These congruent bases are parallel. The shape of each pair of bases can be any polygon. The shape of the base determines the name of the prism. The word "base" does not mean the bottom of the figure. In each figure below, the bases are the front and back of the figure. However, the figures can be turned so that the bases are in different positions.

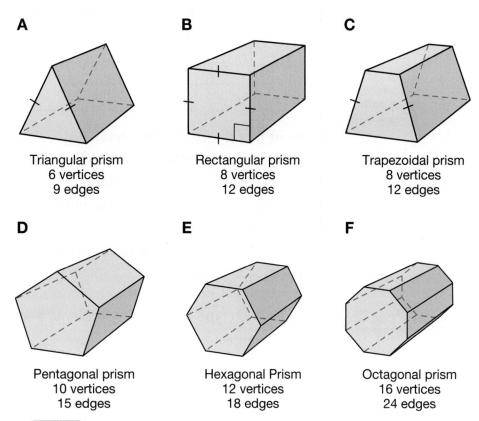

A

Triangular prism
6 vertices
9 edges

B

Rectangular prism
8 vertices
12 edges

C

Trapezoidal prism
8 vertices
12 edges

D

Pentagonal prism
10 vertices
15 edges

E

Hexagonal Prism
12 vertices
18 edges

F

Octagonal prism
16 vertices
24 edges

(**Analyze**) What shape is each pair of bases in prisms **A–F?**
What shape are the faces that are not bases?

Example 1

Which prisms A–F have parallel rectangular faces?

In figure **B,** if we consider the front and back of the rectangular prism as bases, then we see two other pairs of parallel rectangular faces. That is, the top and the bottom faces are parallel, and the left and right faces are also parallel. Also notice that opposite rectangular faces in figures **C, E,** and **F** are parallel. Any prism whose bases have parallel sides has parallel rectangular faces.

(**Conclude**) Do all regular polygons have parallel sides? Why or why not?

Example 2

Which prisms A–F have congruent rectangular faces?

Prism A has 2 congruent rectangular faces since two sides of the triangular base are the same length.

Prism B has four congruent rectangular faces since its bases are squares.

Prism C has 2 congruent rectangular faces since its bases are trapezoids with two sides the same length.

Figures D, E, and F have all congruent rectangular faces since the bases are regular polygons.

> **Classify** Which prism has bases that are rhombuses?

Example 3

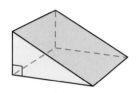

Look at this figure. Does it have any perpendicular rectangular faces?

Notice that two sides of the triangular bases are perpendicular. Therefore, two rectangle faces are also perpendicular to each other.

> **Verify** Which prisms **A–F** have rectangular faces that are perpendicular to each other? Explain why.

Lesson Practice Name each type of prism in problems **a–c.**

a.

b.

c.
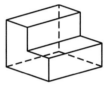

d. The prism in problem **a** has how many rectangular faces that are perpendicular to each other?

e. A rectangular prism has how many pairs of parallel faces?

f. The prism in problem **c** has how many edges?

g. **Explain** Is this figure a prism? Explain your answer.

Written Practice *Distributed and Integrated*

1. **Justify** Tessa made 70 photocopies. If she paid 6¢ per copy and
(93) the total tax was 25¢, how much change should she have gotten back from a $5 bill? Is your answer reasonable? Why or why not?

2. a. What is the area of this square?
(20, Inv. 3)
b. What is the perimeter of the square?

6 cm

*3. Use the information below to answer parts **a** and **b**.
(93)

Walker has $9. Dembe has twice as much money as Walker. Chris has $6 more than Dembe.

    a. How much money does Dembe have?

    b. How much money does Chris have?

4. Use this table to answer the questions that follow:
(25, 93, Inv. 10)

Number of Dumplings	12	24	36	48	60
Number of Dozens	1	2	3	4	5

    a. (Generalize) Write a rule that describes the relationship of the data.

    b. (Predict) How many dumplings is 12 dozen dumplings?

5. (Analyze) There are 40 quarters in a roll of quarters. What is the value of 2 rolls of quarters?
(93)

6. (Estimate) Lucio estimated that the exact quotient of 1754 divided by 9 was close to 20. Did Lucio make a reasonable estimate? Explain why or why not.
(91)

*7. Write the reduced form of each fraction:
(106)

    a. $\frac{2}{12}$ **b.** $\frac{6}{8}$ **c.** $\frac{3}{9}$

8. (Conclude) The three runners wore black, red, and green T-shirts. The runner wearing green finished one place ahead of the runner wearing black, and the runner wearing red was not last. Who finished first? Draw a diagram to solve this problem.
(76, 93)

*9. (Explain) Reduce the fraction $\frac{6}{8}$. Explain your thinking.
(106)

*10. (List) **a.** Write 50 as the product of two factors as many ways as possible.
(55, 56, 105)

    b. Write 50 as the product of prime factors.

*11. **Classify** Name this prism.
(108)

   a. How many vertices does it have?

   b. How many edges does it have?

   c. How many faces does it have?

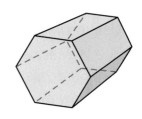

12. 4.62 + 16.7 + 9.8
(45)

13. 14.62 − (6.3 − 2.37)
(45)

*14. $\frac{3}{5} + \frac{1}{5}$
(100)

*15. $16 + 3\frac{3}{4}$
(100)

*16. $1\frac{2}{3} + 3\frac{1}{3}$
(98, 100)

*17. $\frac{2}{5} + \frac{3}{5}$
(98, 100)

*18. $7\frac{4}{5} + 7\frac{1}{5}$
(98, 100)

*19. $6\frac{1}{3} + 3\frac{1}{3}$
(100)

*20. 372 × 39
(107)

*21. 47 × 142
(107)

*22. $375 \times \sqrt{36}$
(Inv. 3, 59)

*23. Estimate the area of this circle. Each small square
(RF14) represents one square centimeter.

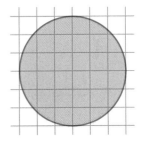

24. 8y = 4832
(34, 83)

25. $\frac{2840}{2^3}$
(65, 79)

*26. $3\overline{)963}$
(79)

27. **Represent** Which arrow could be pointing to 427,063?
(Inv. 2, 27)

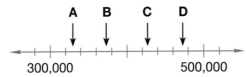

*28. If the length of each side of a square is $1\frac{1}{4}$ inches, then what is the
(20, 100) perimeter of the square?

29. What is the geometric shape of a volleyball?
(96)

*30. Use the Distributive Property to multiply:
(102)

$$5(20 + 6)$$

California Mathematics Content Standards
MG 3.0, 3.1 Identify lines that are parallel and
perpendicular.
MG 3.0, 3.6 Visualize, describe, and make models of
geometric solids (e.g., prisms, pyramids)
in terms of the number and shape of
faces, edges, and vertices; interpret
two-dimensional representations of
three-dimensional objects; and draw
patterns (of faces) for a solid that, when
cut and folded, will make a model of the
solid.

• Constructing Pyramids

facts   Power Up B

**mental
math**

Find each fraction of 40 in problems **a–c.**

a. Fractional Part: $\frac{1}{2}$ of 40

b. Fractional Part: $\frac{1}{4}$ of 40

c. Fractional Part: $\frac{1}{10}$ of 40

d. Money: Shelly gave the clerk a $10 bill for a half gallon
of milk that cost $1.95. How much change should she
receive?

e. Time: Rashid was born on a Monday in April 2000. On what
day of the week was his first birthday?

f. Estimation: Estimate the area of
the rectangle shown at right.

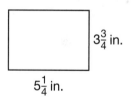

$3\frac{3}{4}$ in.

$5\frac{1}{4}$ in.

g. Calculation: $\sqrt{64}$, $- 3$, $\times 7$, $- 3$, $\div 8$

h. Simplify: $39 - g^2$ when $g = 3$

**problem
solving**

Choose an appropriate problem-solving strategy to solve this
problem. There are four parking spaces (1, 2, 3, and 4) in the row
nearest to the entrance of the building. Suppose only two of the
four parking spaces are filled. What are the combinations of two
parking spaces that could have cars in them?

Math Language

A **plane** is a 2-dimensional, flat surface that never ends. Lines and plane figures are found on planes.

Recall from Lesson 50 that geometric shapes such as triangles, rectangles, and circles have two dimensions—length and width—but they do not have depth. These kinds of figures occupy area, but they do not take up space. We call shapes such as these plane figures because they are confined to a plane.

square

triangle

circle

Shapes that take up space are *geometric solids* such as cubes, pyramids, and cones. Geometric solids have three dimensions: length, width, and depth. Sometimes we simply call these shapes solids. Solids are not confined to a plane, so to draw them we try to create an optical illusion to suggest their shape.

cube

pyramid

cone

In Lesson 97 we studied models of rectangular prisms and triangular prisms. In this lesson we will study models of pyramids.

Math Language

A pyramid is a three-dimensional solid with one base that can be any polygon. The base of a pyramid is not a face.

Constructing Models of Pyramids

Materials needed:
- **Lesson Activity 31**
- scissors
- glue or tape

Cut out the patterns for the pyramids. The shaded parts of each pattern are tabs to help hold the figures together. Fold the paper along the edges before you glue or tape the seams. You might want to work with a partner as you construct the models. Refer to the models to answer the following questions.

Refer to the pyramid with a square base at right to answer problems **a–d**.

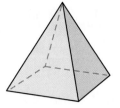

  a. How many faces does the pyramid have, and what are their shapes?

  b. Does the pyramid have any parallel faces?

  c. Does the pyramid have any parallel or perpendicular edges? Explain.

  d. In the pyramid above, what types of angles are formed by the intersecting edges?

Refer to the pyramid with the triangular base at right to answer problems **e–h**.

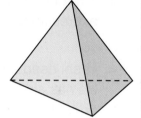

  e. How many faces does the pyramid have and what are their shapes?

  f. Does the pyramid have any parallel faces?

  g. Does the pyramid have any parallel or perpendicular edges?

  h. In the pyramid above, what types of angles are formed by intersecting edges?

Lesson Practice

a. **Verify** Which of these nets can be folded to form a pyramid?

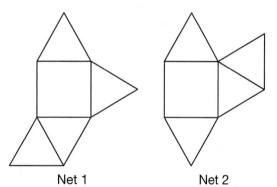

Net 1   Net 2

b. **Model** Use grid paper to draw a net that will fold into a pyramid. Cut it out and fold it up to check your drawing.

1. If a can of soup costs $1.50 and serves 3 people, how much would it
(93) cost to serve soup to 12 people?

*2. **Model** Predict whether this net will fold into a triangular pyramid.
(109) Then draw a net on paper and cut it out to see if your prediction was
correct.

3. What number is eight less than the product of nine and ten? Write an
(63) expression.

4. Yoshi needs to learn 306 new words for the regional spelling bee.
(94) He has already memorized $\frac{2}{3}$ of the new words. How many words
does Yoshi still need to memorize? Draw a picture to illustrate the
problem.

5. a. Find the length of this line segment in centimeters.
(42)

   b. Find the length of the segment in millimeters.

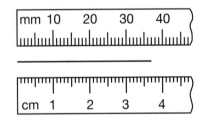

6. **Represent** Use words to write 356,420.
(27)

7. **Represent** Which arrow could be pointing to 356,420?
(Inv. 2,
27)

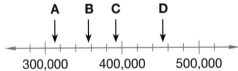

*8. Write the reduced form of each fraction:
(106)

   a. $\frac{2}{6}$ **b.** $\frac{6}{9}$ **c.** $\frac{9}{16}$

*9. a. There were 40 workers on the job. Of those workers, 10 had worked
(Inv. 4, 106) overtime. What fraction of the workers had worked overtime?
(Remember to reduce the fraction.)

b. What decimal part of the workers had worked overtime?

*10. (**Estimate**) Round 14,563,900 to the nearest million.
(104)

*11. (**Explain**) Reduce $\frac{10}{15}$. Explain your reasoning.
(105)

12. (**Conclude**) Jamar received $10 for his tenth birthday. Each year after
(11, 93) that, he received $1 more than he did on his previous birthday. He
saved all his birthday money. In all, how much birthday money did
Jamar have on his fifteenth birthday?

*13. (**Analyze**) Every morning Marta walks $2\frac{1}{2}$ miles. How many miles does
(100) Marta walk in two mornings?

14. $9.36 - (4.37 - 3.8)$
(45)

15. $24.32 - (8.61 + 12.5)$
(45)

*16. $5\frac{5}{8} + 3\frac{3}{8}$
(98, 100)

*17. $6\frac{3}{10} + 1\frac{2}{10}$
(100, 106)

*18. $8\frac{2}{3} - 5\frac{1}{3}$
(100)

*19. $4\frac{3}{4} - 2\frac{1}{4}$
(100, 106)

*20. 125×16
(107)

*21. $12 \times \$1.50$
(107)

22. $6m = 3642$
(34, 83)

23. $\$125 \div 5$
(69)

*24. $4\overline{)645}$
(79)

25. $3m = 6^2$
(25, 34, 65)

26. (**Evaluate**) If n is 16, then what does $3n$ equal?
(38, 63)

27. Dion's temperature is 99.8°F. Normal body temperature is about
(21, 45) 98.6°F. Dion's temperature is how many degrees above normal body
temperature?

*28. Estimate the perimeter and area of this piece of land. Each
(RF14) small square represents one square mile.

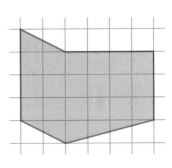

*29. (Analyze) Write the dimensions of a rectangle with a perimeter of 80 yd
(31) and an area of 400 yd^2.

*30. (Conclude) Use this figure to answer parts **a–d**.
(108)
    a. Name this figure.

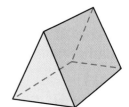

    b. How many vertices does it have?

    c. How many edges does it have?

    d. How many faces does it have?

Early Finishers

Real-World Connection

What is the length of the line segment on the grid? Explain how you found your answer.

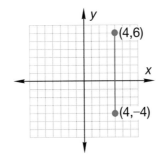

California Mathematics Content Standards
SDAP 2.0, 2.2 Express outcomes of experimental probability situations verbally and numerically (e.g., 3 out of 4; $\frac{3}{4}$).

• Simple Probability

facts   Power Up B

mental math   Find each fraction of 100 in problems **a–c.**

a. Fractional Part: $\frac{1}{2}$ of 100

b. Fractional Part: $\frac{1}{4}$ of 100

c. Fractional Part: $\frac{1}{10}$ of 100

d. Number Sense: 5×46

e. Money: Doug purchased socks for $4.37 and a hairbrush for $2.98. How much did he spend?

f. Estimation: Estimate the area of the rectangle shown at right.

g. Calculation: 12×3, $\sqrt{}$, $\div 2$, $\div 3$

h. Simplify: $4 + h^2$ when $h = 4$

$5\frac{3}{4}$ in. $2\frac{3}{4}$ in.

problem solving   Choose an appropriate problem-solving strategy to solve this problem. Using at least one of each coin from a penny through a half-dollar, which nine coins would be needed to make exactly 99¢?

New Concept

Probability is a measure of how likely it is that an event (or combination of events) will occur. Probabilities are numbers between 0 and 1. An event that is **certain** to happen has a probability of 1. An event that is *impossible* has a probability of 0. If an event is uncertain to occur, then its probability is a fraction between 0 and 1.

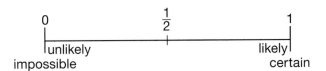

0          $\frac{1}{2}$ 1

unlikely   likely

impossible   certain

The more likely an event, the closer its probability is to 1. The more unlikely an event, the closer its probability is to 0.

Suppose this nickel was tossed once. How might the coin land? Will it land heads up? Or will it land tails up?

The different ways the coin might land represent the outcomes of the toss. An **outcome** is a result. For a coin toss, there are two outcomes. It is just as likely for the coin to land heads up as it is to land tails up.

The probability of the coin landing heads up is "1 out of 2." The probability of the coin landing tails up is "1 out of 2."

When probability is expressed as a fraction, we can read the fraction different ways. For example, if the probability of an event is $\frac{1}{2}$, we can read the probability as "1 out of 2" or "one half."

Example 1

Describe the probability of each outcome in a–c.

a. **What is the probability of the spinner stopping on E?**

b. **What is the probability of the spinner stopping on a letter?**

c. **What is the probability of the spinner stopping on A?**

a. Since there is no letter E on the spinner, it is impossible to land on the letter E. The probability is **0.**

b. Each section has a letter on it, so it is certain that the spinner will land on a letter. The probability is **1.**

c. Since the circle is divided into three equal parts, it is just as likely for the spinner to land on each of the letters. The probability of landing on A is **1 out of 3 or $\frac{1}{3}$.**

Example 2

Thinking Skills

Analyze

If two marbles are chosen from the bag each time, what are the possible outcomes? Show your answer using a table.

A brown paper bag contains black, blue, and white marbles that are the same size. The colors and number of each color are shown below. If one marble is chosen without looking, which color is most likely to be chosen?

4 black   3 white   5 blue

Since there are more yellow marbles than red or green marbles in the bag, **it is most likely that a blue marble will be chosen.**

Connect What color is least likely to be chosen?

Lesson Practice

Analyze The spinner at the right is spun once.

a. Describe the probability of the spinner landing on B.

b. What is the probability of the spinner landing on C? Write a fraction to show your answer.

c. What is the probability of the spinner landing on P?

d. Five pennies are placed in a bag. The year in which each penny was minted is shown below:

2006   1999   1976   1998   2007

Suppose you reach into the bag and choose a penny without looking. What fraction describes the probability that you will choose the 1976 penny?

e. If a dime is dropped on the floor, what is the probability that it will land heads up?

f. Chad printed his first name on a piece of paper. Then he cut out his name, cut apart the letters, and placed the letters in a bag. If Chad pulls one letter out of the bag, what is the probability that the letter will be a C?

g. Read problem **f** again. Is it likely or unlikely that the letter Chad picks will be a consonant?

Written Practice *Distributed and Integrated*

1. Evan found 24 seashells. If he gave one fourth of them to his brother,
(94) how many did he keep?

2. Rectangular Park is 2 miles long and 1 mile wide. Gordon ran
(20) around the park twice. How many miles did he run?

2 mi

1 mi

3. If 2 oranges cost 42¢, how much would 8 oranges cost?
(93)

4. a. (**Represent**) Three fourths of the 64 baseball cards showed rookie
(67, 94) players. How many of the baseball cards showed rookie players?
Draw a picture to illustrate the problem.

   b. What fractional part of the baseball cards were not rookie players?

5. Write these numbers in order from greatest to least:
(Inv. 4,
43)
$$7.2 \quad 7\frac{7}{10} \quad 7\frac{3}{10} \quad 7.5$$

6. Multiple Choice Which of these fractions is *not* equivalent to $\frac{1}{2}$?
(98)
$\quad$ **A** $\frac{3}{6}$ $\qquad$ **B** $\frac{5}{10}$ $\qquad$ **C** $\frac{10}{21}$ $\qquad$ **D** $\frac{50}{100}$

*7. Complete each equivalent fraction:
(103)
$\quad$ **a.** $\frac{1}{2} = \frac{?}{12}$ $\qquad$ **b.** $\frac{1}{3} = \frac{?}{12}$ $\qquad$ **c.** $\frac{1}{4} = \frac{?}{12}$

*8. Write the reduced form of each fraction:
(106)
$\quad$ **a.** $\frac{5}{10}$ $\qquad$ **b.** $\frac{8}{15}$ $\qquad$ **c.** $\frac{6}{12}$

9. (**Analyze**) Caleb paid 42¢ for 6 clips and 64¢ for 8 erasers. What was
(93) the cost of each clip and each eraser? What would be the total cost of
10 clips and 20 erasers?

10. (**Conclude**) There were 14 volunteers the first year, 16 volunteers
(64, 93) the second year, and 18 volunteers the third year. If the number of
volunteers continued to increase by 2 each year, how many volunteers
would there be in the tenth year? Explain how you know.

*11. (**Represent**) Write the number 16 as the product of prime factors.
(105)

12. (**Predict**) A standard dot cube is rolled. What is the probability that the
(110) number rolled will be less than seven?

13. $47.14 - (3.63 + 36.3)$ $\qquad\qquad$ **14.** $50.1 + (6.4 - 1.46)$
(9, 45) $\qquad\qquad\qquad\qquad\qquad\qquad\quad$ *(9, 45)*

* 15. $\frac{2}{4} + \frac{1}{4} + \frac{1}{4}$
(98, 100)

* 16. $4\frac{1}{6} + 1\frac{1}{6}$
(100, 106)

* 17. $5\frac{3}{5} + 1\frac{2}{5}$
(100, 106)

* 18. $\frac{5}{6} + \frac{1}{6}$
(100)

* 19. $12\frac{3}{4} - 3\frac{1}{4}$
(100, 106)

* 20. $6\frac{1}{5} - 1\frac{1}{5}$
(100)

* 21. 340×15
(107)

* 22. 26×337
(107)

* 23. 72×251
(107)

24. $\frac{3550}{5}$
(83)

* 25. $432 \div 3$
(79)

26. $9\overline{)5784}$
(79)

* 27. Karen is planning a trip to Los Angeles from Chicago for her vacation.
(RF13, 76) She finds the following two round-trip flight schedules. Use this information to answer parts **a–c.**

Passengers: 1			Price: $246.00	
Flight number	Departure city	Date Time	Arrival city	Date Time
12A	ORD Chicago	7/21 06:11 PM	LAX Los Angeles	7/21 08:21 PM
46	LAX Los Angeles	7/28 06:39 PM	ORD Chicago	7/29 12:29 AM

Passengers: 1			Price: $412.00	
Flight number	Departure city	Date Time	Arrival city	Date Time
24	ORD Chicago	7/21 08:17 AM	LAX Los Angeles	7/21 10:28 AM
142	LAX Los Angeles	7/28 03:28 PM	ORD Chicago	7/28 09:18 PM

a. If Karen wants to arrive in Los Angeles in the morning, how much will she pay for airfare?

b. If Karen chooses the more economical round trip, when is her return flight scheduled to land?

c. Multiple Choice There is a 2-hour time difference between Chicago and Los Angeles. About how long does a flight between those cities last?

A 2 hours   **B** 4 hours   **C** 6 hours   **D** 8 hours

*28. Jenna is playing a board game. She has one dot cube and wants to roll
(110) a 5. What is the probability she rolls a 5 in one roll?

*29. (**Explain**) Is this figure a pyramid? Why or why not?
(109)

*30. (**Analyze**) Use this figure to for parts **a–c.**
(108)
    a. (**Classify**) What is the geometric name for the shape
    of a cereal box?

    b. How many edges does this box have?

    c. Describe the angles.

California Mathematics Content Standards
SDAP 1.0, 1.3 Interpret one-and two-variable data graphs to answer questions about a situation.
SDAP 2.0, 2.1 Represent all possible outcomes for a simple probability situation in an organized way (e.g., tables, grids, tree diagrams).
SDAP 2.0, 2.2 Express outcomes of experimental probability situations verbally and numerically (e.g., 3 out of 4; $\frac{3}{4}$).

Focus on

Probability Experiments

Many board games involve an element of **chance.** This means that when we spin a spinner, roll number cubes, or draw a card from a shuffled deck, we cannot know the outcome (result) of the event ahead of time. However, we can often find how *likely* a particular outcome is. The degree of likelihood of an outcome is called its probability.

Here we show a spinner. The face is divided into six equal parts called **sectors.** Each sector is $\frac{1}{6}$ of the face of the spinner. Assuming the spinner is balanced and fair, then a spin of the arrow can end up with the arrow pointing in any direction. The letter that names the sector where the arrow lands is the outcome of the spin. For the questions that follow, ignore the possibility that the arrow may stop on a line.

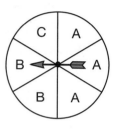

1. If the arrow is spun once, what outcomes are possible?

2. **Explain** On which letter is the arrow most likely to stop, and why?

3. **List** Write the possible outcomes of a spin in order from least likely to most likely.

4. Which outcome of a spin is twice as likely as the outcome C?

5. **Predict** If the arrow is spun many times, then about half the outcomes are likely to be which sector?

6. **Multiple Choice** If the arrow is spun many times, then what fraction of the spins are likely to stop in sector C?

   A $\frac{1}{6}$ **B** $\frac{1}{3}$ **C** $\frac{1}{2}$ **D** $\frac{5}{6}$

7. **Multiple Choice** In 60 spins, about how many times should we expect it to stop in sector C?

   A about 6 times   **B** about 10 times
   C about 20 times   **D** about 30 times

Recall that probability of an outcome can be expressed as a number ranging from 0 to 1. An outcome that cannot happen has a probability of 0. An outcome that is certain to happen has a probability of 1. An outcome that could happen but is not certain to happen is expressed as a fraction between 0 and 1.

Use the spinner at right to answer problems **8–10.**

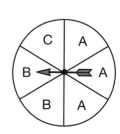

8. **Explain** What is the probability that the arrow will stop in sector D? Why?

9. **Explain** What is the probability that the outcome will be one of the first three letters of the alphabet? Why?

10. What is the probability that the arrow will stop in sector C?

Here we show a standard dot cube.

11. What numbers are represented by the dots on the faces of a dot cube?

12. **Justify** If a dot cube is rolled once, which number is most likely to end up on top? Why?

13. **Multiple Choice** If a dot cube is rolled many times, about how often would we expect to roll a number greater than 3?

    A less than half the time   **B** about half the time

    C more than half the time   **D** none of the time

14. If a dot cube is rolled once, what is the probability of rolling a 7?

15. With one roll of a dot cube, what is the probability of rolling a 1?

16. **Multiple Choice** How would we describe the likelihood of rolling a 6 with one roll of a dot cube?

    A very likely   **B** just as likely to roll a 6 as not to roll a 6

    C unlikely   **D** certain

Probability Experiments

Materials needed:

- **Lesson Activity 32**
- dot cubes

Experiment 1: Work with a partner for this experiment. You and your partner will roll one dot cube 36 times and tally the number of times each face of the dot cube turns up. You will record the results in the Experiment 1 table on **Lesson Activity 32.** (A copy of the table is shown on the next page.)

Predict Before starting the experiment determine all the possible outcomes. For this experiment there are 6 possible outcomes, 1, 2, 3, 4, 5 and 6. Then, predict the number of times each outcome will occur during the experiment. Write your predictions in the column labeled "Prediction."

36 Rolls of One Dot Cube

Outcome	Prediction	Tally	Total Frequency
1			
2			
3			
4			
5			
6			

Now begin rolling the dot cube. Make a tally mark for each roll in the appropriate box in the "Tally" column. When all groups have finished, report your results to the class. As a class, total the groups' tallies for each outcome, and write these totals in the boxes under "Total Frequency."

17. Make a bar graph using the data from your table.

18. (Conclude) What conclusions can you draw from the results of Experiment 1?

19. What was the mode for the outcomes of this experiment?

20. Is it easier to compare data using the bar graph or the table?

Experiment 2: In this experiment you and your group will roll a pair of dot cubes 36 times and tally the outcomes. For each roll the outcome will be the sum of the two numbers that end up on top. You will record your results in the Experiment 2 table on **Lesson Activity 32.**

Form groups so that each group can have two number cubes.

(Predict) Before starting the experiment, predict as a group the number of times each outcome will occur during the experiment. Write your predictions in the column labeled "Prediction."

36 Rolls of Two Dot Cubes

Outcome	Prediction	Tally	Total Frequency
2			
3			
4			
5			
6			
7			
8			
9			
10			
11			
12			

Now begin rolling the dot cubes. Each time you roll a pair of dot cubes, make a tally mark in the appropriate box. When all groups have finished, report your results to the class. As a class, total the groups' tallies for each outcome and record these totals in the "Total Frequency" column.

21. What was the mode of the outcomes? Why?

22. Which outcome(s) occurred least frequently? Why?

23. What conclusions can you draw from the results of Experiment 2?

24.  **Model** What are all the possible combinations you could roll with a sum of 7 as the result? Explain.

Investigate Further

a. Logan brought two hats for a weekend trip: one navy blue and one black. He also brought three shirts: one red, one green, and one yellow. We can use a tree diagram to show all the possible combinations of shirts and hats he can wear.

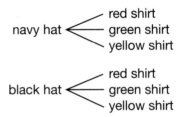

We can count the branches to see all the possible combinations. There are six combinations of hats and shirts.

Represent Logan brought three pairs of shorts: denim, brown, and black. Draw a tree diagram to show all the possible combinations of hats, shirts, and shorts. How many are there?

b. There are two spinners shown below. Spinner A is labeled with letters and Spinner B is labeled with numbers.

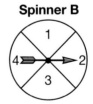

We can show all possible outcomes for the two spinners on a grid. We make a column for each possible letter and pair each possible number with them. We can see that there are 12 possible outcomes.

**Outcomes for
Spinners A and B**

R, 1	S, 1	T, 1
R, 2	S, 2	T, 2
R, 3	S, 3	T, 3
R, 4	S, 4	T, 4

Interpret What is the probability of spinning the letter T and the number 4?

Represent Use the grid to make a tree diagram of all the possible outcomes for the Spinners A and B.

California Mathematics Content Standards

NS **3.0, 3.3** Solve problems involving multiplication of multidigit numbers by two-digit numbers.

MR **3.0, 3.2** Note the method of deriving the solution and demonstrate a conceptual understanding of the derivation by solving similar problems.

MR **3.0, 3.3** Develop generalizations of the results obtained and apply them in other circumstances.

• Multiplying Three-Digit Numbers

facts Power Up C

mental math Find each fraction of 60 in problems **a–c.**

a. Fractional Part: $\frac{1}{3}$ of 60

b. Fractional Part: $\frac{2}{3}$ of 60

c. Fractional Part: $\frac{3}{3}$ of 60

d. Number Sense: 50×46

e. Probability: With one roll of a dot cube, what is the probability of rolling a 4?

f. Estimation: Estimate 49×21.

g. Calculation: $\frac{1}{3}$ of 90, $+ 50$, $+ 1$, $\sqrt{\ }$, $\sqrt{\ }$, $- 2$

h. Simplify: $k^2 + 25$ when $k = 5$

problem solving Choose an appropriate problem-solving strategy to solve this problem. Marco paid a dollar for an item that cost 54¢. He received four coins in change. What four coins did he receive?

New Concept

When we multiply a three-digit number by a multiple of ten, we can find the product by doing only one multiplication instead of two.

$$\begin{array}{r} 243 \\ \times\ 20 \end{array}$$ We can rewrite the problem so the zero "hangs out" to the right. $$\begin{array}{r} 243 \\ \times\quad 20 \end{array}$$

$$\begin{array}{r} 243 \\ \times\quad 20 \\ \hline 4860 \end{array}$$

Multiplying by 243 by 20 is the same as multiplying 243 by 10 and then by 2.

We multiply by 2. ———⌐ ⌐——— We write the zero in the product to show that we multiplied by 10.

The product is **4860.**

Example 1

Multiply: 20 × 306

First we write the factors so the zero "hangs out."

$$\begin{array}{r} 1 \\ 306 \\ \times\quad 20 \\ \hline 6120 \end{array}$$

Step 2
We think 2 × 0 = 0 plus 1 = 1.
We write the 1 in the product.
We multiply by 3.

Step 1
We think 2 × 6 = 12. Write the 2 in the product and write the 1 ten above the 0.

The product is **6120.**

Example 2

Multiply: 24 × 406

Since there are two digits in the multiplier 24, we will have two partial products.

First we multiply by 4 ones. We think 4 × 6 = 24. Write the 4 in the product and write the 2 tens above the 0. Next, we multiply 0 by 4. 4 × 0 = 0 plus 2 = 2. Write the two in the tens place of the product. Then, we multiply the hundreds. 4 × 4 = 16.

$$\begin{array}{r} 1 \\ 2 \\ 406 \\ \times\quad 24 \\ \hline 1624 \\ 812\quad \\ \hline 9744 \end{array}$$

Now we multiply by 2, which is really 20. We think 2 × 6 = 12. Write the 2 in the tens place of the product and write the 1 above the 0. Then we multiply 2 × 0 = 0 plus 1 = 1. Write the 1 in the product. Now we multiply 2 × 4 and write 8 in the product.

Finally, we add the partial products.

The product is **9744.**

Discuss Why did we move one place to the left when we multiplied by the 2 tens in 24?

a. 402 × 30

b. 543 × 40

c. 804 × 60

d. 320 × 50

e. 36 × 115

f. 419 × 63

g.   102
    ×   12

h.   404
    ×   25

i.   125
    ×   16

j.   306
    ×   23

Written Practice
Distributed and Integrated

1. **Explain** Forty-five students are separated into four groups.
(87) The number of students in each group is as equal as possible.
How many students are in the largest group? Explain your
reasoning.

2. Use the formulas to solve parts **a–b.**
(66, 101)
   a. What is the area of this rectangle?

   b. What is the perimeter of this rectangle?

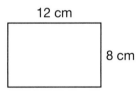
12 cm

8 cm

* **3.** **Represent** Iggy answered $\frac{5}{6}$ of the 90 questions correctly. How many
(94) questions did Iggy answer correctly? Draw a picture to illustrate
the problem.

4. Name the shape of each object:
(96)
   a. roll of paper towels   **b.** baseball

* **5.** Write the reduced form of each fraction:
(106)
   a. $\frac{3}{6}$ **b.** $\frac{5}{15}$ **c.** $\frac{8}{12}$

* **6.** **Analyze** **a.** A box contains marbles that are the same size. There are
(110,
Inv. 11)   3 blue, 7 purple, and 5 red marbles in the box. If one marble is
chosen without looking, which color is least likely to be chosen?

   b. What number is least likely to be rolled on a number cube labeled
1, 2, 3, 4, 5, and 6?

7. Which digit is in the ten-millions place in 328,496,175?
(49)

8. (Analyze) Draw a picture to help you solve this problem:
(33)
Winder is between Atlanta and Athens. It is 73 miles from Athens to Atlanta. It is 23 miles from Winder to Athens. How many miles is it from Winder to Atlanta?

9. Lyle volunteers after school as a tutor. Each afternoon he begins a tutoring session at the time shown on the clock and finishes three quarters of an hour later. What time does each tutoring session end?
(13)

*10. In 1994, Caribou, Maine, had a record high temperature of 59°F in February. In 1955, Caribou's record low temperature in February was −41°F. What is the difference between these two record temperatures?
(21)

11. 4.36 + 12.7 + 10.72
(45)

12. 8.54 − (4.2 − 2.17)
(9, 45)

*13. $\frac{5}{9} + \frac{3}{9}$
(100)

*14. $3\frac{2}{3} + 1\frac{1}{3}$
(98, 100)

15. $4\frac{5}{8} + 1$
(100)

*16. $7\frac{1}{4} + 1\frac{2}{4}$
(100)

*17. $4\frac{4}{9} + 1\frac{1}{9}$
(100)

*18. $\frac{11}{12} + \frac{1}{12}$
(98, 100)

*19. 507 × 60
(107)

*20. 382 × 31
(107)

21. 505 × 22
(111)

22. $\frac{3731}{7}$
(79)

23. $9\overline{)5432}$
(79)

*24. $6\overline{)548}$
(72)

25. (Predict) The first five square numbers are 1, 4, 9, 16, and 25.
(Inv. 3)

What is the eighth term of this sequence? Write an equation to support your answer.

*26. In the year 2006 the population of the United States reached three hundred million. Use digits to write the number in standard form.
(49)

27. a. Multiple Choice Dacus built a square frame using
(17, 90)   two-by-fours, but when he leaned against it, the frame
           shifted to this shape. What word does *not* name this
           shape?

    A quadrilateral   **B** parallelogram

    C rhombus   **D** trapezoid

  b. Describe the angles.

  c. Describe the sides.

28. If the perimeter of a square is 6 centimeters, then each side is how
(20, 42) many millimeters long?

29. A cube has how many more vertices than this pyramid?
(96, 108
109)

*30. Three teaspoons are equal to one tablespoon. Use this relationship for
(Inv. 7,
Inv. 9, parts **a–c.**
Inv. 10)

  a. (**Generalize**) Write a formula for changing any number of
     tablespoons to teaspoons. Use *t* for teaspoons and *b* for
     tablespoons.

  b. (**Evaluate**) Write a set of ordered pairs for 1, 2, and 3
     tablespoons.

  c. (**Represent**) Graph the ordered pairs and draw a line to represent
     the equation.

LESSON

112

California Mathematics Content Standards

NS 1.0, 1.5 Explain different interpretations of fractions, for example, parts of a whole, parts of a set, and division of whole numbers by whole numbers; explain equivalents of fractions (see Standard 4.0).

NS 1.0, 1.7 Write the fraction represented by a drawing of parts of a figure; represent a given fraction by using drawings; and relate a fraction to a simple decimal on a number line.

• Simplifying Fraction Answers

facts	Power Up G
mental math	Find each fraction of 60 in problems **a–c**.

    a. Fractional Part: $\frac{1}{4}$ of 60

    b. Fractional Part: $\frac{2}{4}$ of 60

    c. Fractional Part: $\frac{3}{4}$ of 60

    d. Number Sense: 30×12

    e. Money: Taima had $10.00. Then she spent $5.63 on a journal. How much money does she have left?

    f. Estimation: Eight bottles of laundry detergent cost $40.32. Round that amount to the nearest dollar and then divide by 8 to estimate the cost per bottle.

    g. Calculation: $\frac{1}{2}$ of 24, ÷ 6, square the number, + 8, × 2

    h. Simplify: $m^2 - 36$ when $m = 6$

problem solving

Choose an appropriate problem-solving strategy to solve this problem. Find the next five terms in this sequence. Then describe the sequence in words.

$$\frac{1}{2}, \frac{2}{4}, \frac{3}{6}, \frac{4}{8}, \underline{\quad}, \underline{\quad}, \underline{\quad}, \underline{\quad}, \underline{\quad}, \ldots$$

New Concept

We often write answers to math problems in the simplest form possible. If an answer contains a fraction, there are two procedures that we usually follow.

    1. We write improper fractions as mixed numbers (or whole numbers).

    2. We reduce fractions when possible.

Example 1

Thinking Skills

Justify

Explain why $\frac{4}{3} = 1\frac{1}{3}$.

Add: $\frac{2}{3} + \frac{2}{3}$

We add the fractions and get the sum $\frac{4}{3}$. Notice that $\frac{4}{3}$ is an improper fraction. We take the extra step of changing $\frac{4}{3}$ to the mixed number $1\frac{1}{3}$.

$$\frac{2}{3} + \frac{2}{3} = \frac{4}{3}$$

$$\frac{4}{3} = 1\frac{1}{3}$$

Example 2

Subtract: $\frac{3}{4} - \frac{1}{4}$

We subtract and get the difference $\frac{2}{4}$. Notice that $\frac{2}{4}$ can be reduced. We take the extra step of reducing $\frac{2}{4}$ to $\frac{1}{2}$.

$$\frac{3}{4} - \frac{1}{4} = \frac{2}{4}$$

$$\frac{2}{4} = \frac{1}{2}$$

Example 3

Nicholas exercises each day by walking. The route he walks each morning is $3\frac{1}{3}$ miles long, and the route he walks each evening is $4\frac{2}{3}$ miles long. Altogether, how many miles does Nicholas walk each day?

We add the mixed numbers and get the sum $7\frac{3}{3}$. Notice that $\frac{3}{3}$ is an improper fraction equal to 1. So $7\frac{3}{3} = 7 + 1$, which is 8. Nicholas walks **8 miles** altogether.

$$3\frac{1}{3} + 4\frac{2}{3} = 7\frac{3}{3}$$

$$7\frac{3}{3} = 8$$

Example 4

Add: $5\frac{3}{5} + 6\frac{4}{5}$

We add the mixed numbers and get $11\frac{7}{5}$. Notice that $\frac{7}{5}$ is an improper fraction that can be changed to $1\frac{2}{5}$. So $11\frac{7}{5}$ equals $11 + 1\frac{2}{5}$, which is **$12\frac{2}{5}$**.

$$5\frac{3}{5} + 6\frac{4}{5} = 11\frac{7}{5}$$

$$11\frac{7}{5} = 12\frac{2}{5}$$

Example 5

Thinking Skills

Represent

Draw a picture to show that $\frac{2}{8} = \frac{1}{4}$.

A piece of fabric $1\frac{3}{8}$ yards in length was cut from a bolt of fabric that measured $6\frac{5}{8}$ yards long. How long is the piece of fabric left on the bolt?

We subtract and get $5\frac{2}{8}$. Notice that $\frac{2}{8}$ can be reduced, so we reduce $\frac{2}{8}$ to $\frac{1}{4}$ and get $5\frac{1}{4}$. The length of the fabric is **$5\frac{1}{4}$ yards.**

$$6\frac{5}{8} - 1\frac{3}{8} = 5\frac{2}{8}$$

$$5\frac{2}{8} = 5\frac{1}{4}$$

Lesson Practice   Simplify the answer to each sum or difference:

a. $\dfrac{4}{5} + \dfrac{4}{5}$ **b.** $\dfrac{5}{6} - \dfrac{1}{6}$ **c.** $3\dfrac{2}{3} + 1\dfrac{2}{3}$

d. $5\dfrac{1}{4} + 6\dfrac{3}{4}$ **e.** $7\dfrac{7}{8} - 1\dfrac{1}{8}$ **f.** $5\dfrac{3}{5} + 1\dfrac{3}{5}$

g. $4\dfrac{3}{10} - 3\dfrac{1}{10}$ **h.** $\dfrac{3}{4} + \dfrac{3}{4}$ **i.** $\dfrac{5}{8} - \dfrac{1}{8}$

Written Practice   *Distributed and Integrated*

*1. **(Interpret)** Use the information in the graph
(21, Inv. 5) to answer parts **a–c.**

    a. On which day was the temperature the highest?

    b. What was the high temperature on Tuesday?

    c. From Monday to Wednesday, the temperature went up how many degrees?

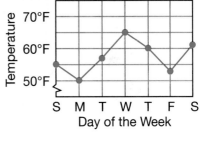

High Temperatures for the Week

2. Use a formula for parts **a** and **b.**
(66, 88, 101)   **a.** What is the perimeter of this rectangle?

    b. What is the area of the rectangle?

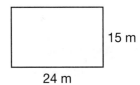

15 m

24 m

*3. **(Analyze)** Simplify each expression. Remember to use the order of
(63)   operations.

    a. $3 \times 5 + (12 \div 6)$ **b.** $15 \div 3 - (3 + 2)$

    c. $14 + (8 - 2) \times 5$ **d.** $6 \times (20 - 12) - 5$

*4. **(Represent)** What fractional part of the months of the year begins with J?
(74, 106)   Reduce the fraction if necessary.

5. There are 52 cards in a deck. Four of the cards are aces. What is the
(110, Inv. 11)   probability of drawing an ace from a full deck of cards?

6. **Classify** Name each shape:
(96)

    a.   **b.**   **c.**

*7. Write the reduced form of each fraction:
(106)

    a. $\dfrac{6}{8}$ **b.** $\dfrac{4}{9}$ **c.** $\dfrac{4}{16}$

*8. **Evaluate** This table represents the equation $y = 3x + 1$ and shows the
(64, 93) values of y when x is 4. What is y when x is 5?

x	y
3	10
4	13
5	?

*9. **Represent** Use words to write the number 27386415.
(49)

10. **Represent** Point W stands for what number on this number line?
(Inv. 2)

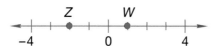

11. **Represent** Draw two parallel segments that are one inch long and
(90) one inch apart. Then make a quadrilateral by drawing two more parallel segments. What type of quadrilateral did you draw?

*12. $4\dfrac{4}{5} + 3\dfrac{3}{5}$ ***13.** $5\dfrac{1}{6} + 1\dfrac{2}{6}$ ***14.** $7\dfrac{3}{4} + \dfrac{1}{4}$
(112)   (112)   (112)

*15. $6\overline{)508}$ ***16.** $4\overline{)3018}$ **17.** $5\dfrac{3}{8} + 5\dfrac{1}{8}$
(72)   (79)   (112)

*18. 25×408 **19.** 702×36 **20.** 147×54
(111)   (107)   (107)

21. $8\overline{)5766}$ ***22.** $2\overline{)440}$
(83)   (83)

23. $4.75 + 16.14 + 10.9$ **24.** $18.4 - (4.32 - 2.6)$
(45)   (9, 45)

*25. **Estimate** In the year 2000 the population of the state of California
(104) was 33,871,648. Round that number to the nearest million.

*26. **Estimate** Round 297,576,320 to the nearest hundred million.
(104)

27. On Jahzara's first nine games she earned these scores:
(Inv. 6, 95)
$$90, 95, 80, 85, 100, 95, 75, 95, 90$$
Use this information to answer parts **a** and **b**.

    a. What is the median and range of Jahzara's scores?

    b. What is the mode of Jahzara's scores?

28. Write these numbers in order from least to greatest:
(43, Inv. 8)

$$5\frac{11}{100} \qquad 5.67 \qquad 5.02 \qquad 5\frac{83}{100}$$

29. Maranie wanted to divide 57 buttons into 9 groups. How many groups
(54) will Maranie have? Will there be any buttons left over?

*30. a. **Explain** Which prism has more vertices? Explain why?
(96, 108, 109)

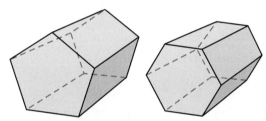

    b. **Justify** A pyramid can have a base that is any polygon. How
many vertices does a pentagonal pyramid have?

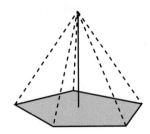

California Mathematics Content Standards

NS 1.0, 1.5 Explain different interpretations of fractions, for example, parts of a whole, parts of a set, and division of whole numbers by whole numbers; explain equivalents of fractions (see Standard 4.0).

MR 2.0, 2.4 Express the solution clearly and logically by using the appropriate mathematical notation and terms and clear language; support solutions with evidence in both verbal and symbolic work.

• Renaming Fractions and Common Denominators

facts   Power Up I

mental math   An odd number can be written as an even number plus 1. For example, 9 is 8 + 1. So half of 9 is half of 8 plus half of 1, which is $4 + \frac{1}{2}$, or $4\frac{1}{2}$. Use this strategy to find half of each odd number in problems **a–d.**

    a. Fractional Part: 7

    b. Fractional Part: 11

    c. Fractional Part: 21

    d. Fractional Part: 33

    e. Probability: If the chance of rain is $\frac{3}{10}$, what is the chance that it will not rain?

    f. Estimation: Uzuri's mother filled the car with gasoline, which cost $33.43. Then her mother bought snacks for $4.48. Estimate the total cost.

    g. Calculation: $\frac{1}{2}$ of 100, $- 1$, $\sqrt{}$, $+ 2$, $\sqrt{}$, $+ 1$, $\sqrt{}$

    h. Simplify: $n^2 + 1$ when $n = 1$

problem solving   Choose an appropriate problem-solving strategy to solve this problem. The numbers 1, 8, and 27 begin the sequence below. (Notice that $1 = 1^3$, $8 = 2^3$, and $27 = 3^3$.) Find the next three numbers in the sequence.

$$1, 8, 27, \underline{}, \underline{}, \underline{}, \ldots$$

Remember that when we multiply a fraction by a fraction name for 1, the result is an equivalent fraction. For example, if we multiply $\frac{1}{2}$ by $\frac{2}{2}$, we get $\frac{2}{4}$. The fractions $\frac{1}{2}$ and $\frac{2}{4}$ are equivalent fractions because they have the same value.

$$\frac{1}{2} \times \frac{2}{2} = \frac{2}{4}$$

Sometimes we must choose a particular multiplier that is equal to 1.

Example 1

Find the equivalent fraction for $\frac{1}{4}$ whose denominator is 12.

To change 4 to 12, we must multiply by 3. So we multiply $\frac{1}{4}$ by $\frac{3}{3}$.

$$\frac{1}{4} \times \frac{3}{3} = \frac{3}{12}$$

The fraction $\frac{1}{4}$ is equivalent to $\frac{3}{12}$.

Example 2

Complete the equivalent fraction: $\frac{2}{3} = \frac{?}{15}$

The denominator changed from 3 to 15. Since the denominator was multiplied by 5, the correct multiplier is $\frac{5}{5}$.

$$\frac{2}{3} \times \frac{5}{5} = \frac{10}{15}$$

Thus, the missing numerator of the equivalent fraction is **10.**

Two or more fractions have **common denominators** if their denominators are equal.

$$\frac{3}{8} \qquad \frac{5}{8} \qquad\qquad\qquad \frac{3}{8} \qquad \frac{5}{9}$$

These two fractions have common denominators.

These two fractions do *not* have common denominators.

We will use common denominators to rename fractions whose denominators are not equal.

Example 3

Rename $\frac{2}{3}$ and $\frac{3}{4}$ so that they have a common denominator of 12.

To rename a fraction, we multiply it by a fraction name for 1. To change the denominator of $\frac{2}{3}$ to 12, we multiply $\frac{2}{3}$ by $\frac{4}{4}$. To change the denominator of $\frac{3}{4}$ to 12, we multiply $\frac{3}{4}$ by $\frac{3}{3}$.

$$\frac{2}{3} \times \frac{4}{4} = \frac{8}{12} \qquad\qquad \frac{3}{4} \times \frac{3}{3} = \frac{9}{12}$$

$$\frac{2}{3} = \frac{8}{12} \qquad\qquad\qquad \frac{3}{4} = \frac{9}{12}$$

Example 4

Math Language

One way to find a *common denominator* is to multiply the denominators.

$$3 \times 4 = 12$$

When we multiply two numbers, each number is a factor of the product.

Rename $\frac{1}{2}$ and $\frac{1}{3}$ so that they have a common denominator.

This time we need to find a common denominator before we can rename the fractions. The denominators are 2 and 3. The product of 2 and 3 is 6, so 6 is a common denominator.

To get denominators of 6, we multiply $\frac{1}{2}$ by $\frac{3}{3}$, and we multiply $\frac{1}{3}$ by $\frac{2}{2}$.

$$\frac{1}{2} \times \frac{3}{3} = \frac{3}{6} \qquad\qquad \frac{1}{3} \times \frac{2}{2} = \frac{2}{6}$$

$$\frac{1}{2} = \frac{3}{6} \qquad\qquad\qquad \frac{1}{3} = \frac{2}{6}$$

Lesson Practice Complete each equivalent fraction:

a. $\frac{1}{4} = \frac{?}{12}$ **b.** $\frac{2}{3} = \frac{?}{9}$ **c.** $\frac{1}{4} = \frac{?}{8}$

d. Rename $\frac{1}{2}$ and $\frac{1}{5}$ so that they have a common denominator of 10.

e. Rename $\frac{1}{2}$ and $\frac{5}{6}$ so that they have a common denominator of 12.

Rename each of these fractions with a denominator of 12.

f. $\frac{2}{3}$ **g.** $\frac{3}{4}$ **h.** $\frac{1}{6}$

Written Practice *Distributed and Integrated*

1. Zuna used 1-foot-square floor tiles to cover the floor of a room 15 feet
(Inv. 3, 88) long and 12 feet wide. How many floor tiles did she use?

2. **a.** What is the perimeter of this triangle?
(20, 45, 81)

    b. Is this triangle equilateral, isosceles, or scalene?

1.2 cm   1.9 cm

2.2 cm

*3. **Represent** Elsa found that $\frac{3}{8}$ of the 32 pencils in the room had no
(94) erasers. How many pencils had no erasers? Draw a picture to illustrate the problem.

*4. **Estimate** Estimate the product of 75 × 75. Explain your reasoning.
(91)

*5. Complete each equivalent fraction:
(103)

    a. $\frac{3}{5} = \frac{?}{10}$ **b.** $\frac{2}{3} = \frac{?}{9}$ **c.** $\frac{1}{4} = \frac{?}{8}$

*6. Write these fractions with a common denominator by multiplying the
(113) denominators. Which fraction is greater?

$$\frac{4}{5} \text{ and } \frac{5}{6}$$

7. Fausta bought 2 DVDs priced at $21.95 each and 2 CDs priced at
(28, 51, 101) $14.99 each. The tax was $4.62. What was the total cost of the items? Explain how you found your answer.

8. Roger drove 285 miles in 5 hours. What was his average speed in miles
(60) per hour?

9. **Multiple Choice** Which of these fractions is *not* equivalent to $\frac{1}{2}$?
(98, 103)

  A $\frac{4}{8}$ **B** $\frac{11}{22}$ **C** $\frac{15}{30}$ **D** $\frac{12}{25}$

*10. Write the reduced form of each fraction:
(106)

    a. $\frac{8}{10}$ **b.** $\frac{6}{15}$ **c.** $\frac{8}{16}$

*11. **Represent** Use words to write the number 123415720 then round it
(104) to the nearest hundred thousand.

12. 8.3 + 4.72 + 0.6 + 12.1   **13.** 17.42 − (6.7 −1.23)
(45)   (9, 45)

*14. $3\frac{3}{8} + 3\frac{3}{8}$ ***15.** $4\frac{1}{6} + 2\frac{1}{6}$ ***16.** $1\frac{1}{6} + \frac{1}{6}$
(112)   (112)   (112)

*17. $5\frac{5}{6} - 1\frac{1}{6}$
(112)

*18. $\frac{2}{8} - \frac{1}{8}$
(100)

*19. $1\frac{4}{6} - \frac{1}{6}$
(112)

*20. 87×16
(88)

*21. 49×340
(107)

*22. 504×30
(107)

23. $\$35.40 \div 6$
(83)

24. $8\overline{)5784}$
(79)

25. $7\overline{)2385}$
(83)

26. $3\overline{)312}$
(83)

*27. $8\overline{)450}$
(72)

*28. $5\overline{)450}$
(75)

29. (Predict) What is the probability of drawing a heart from a full deck
(110, Inv. 11) of cards? (*Hint:* There are 13 hearts in a deck.)

*30. (Represent) Draw a rectangle that is 5 cm long and 2 cm wide, and
(18, 19, 42) divide the rectangle into square centimeters. Then shade $\frac{3}{10}$ of the
rectangle.

*Real-World
Connection*

Estimate the area of the pentagon. Each small square represents one
square inch.

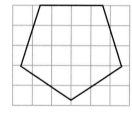

California Mathematics Content Standards

AF 1.0, 1.4 Use and interpret formulas (e.g., area = length × width or $A = lw$) to answer questions about quantities and their relationships.

MG 1.0, 1.1 Measure the area of rectangular shapes by using appropriate units, such as square centimeter (cm²), square meter (m²), square kilometer (km²), square inch (in²), square yard (yd²), or square mile (mi²).

MG 1.0, 1.4 Understand and use formulas to solve problems involving perimeters and areas of rectangles and squares. Use those formulas to find the areas of more complex figures by dividing the figures into basic shapes.

• Perimeter and Area of Complex Figures

facts   Power Up H

mental math

a. Fractional Part: $\frac{1}{4}$ of 24

b. Fractional Part: $\frac{1}{2}$ of 24

c. Fractional Part: $\frac{3}{4}$ of 24

d. Number Sense: 20×250

e. Measurement: The half-gallon container is half full. How many quarts of liquid are in the container?

f. Estimation: Each square folding table is 122 cm on each side. Estimate the total length of 4 folding tables if they are lined up in a row.

g. Calculation: $6^2 - 6$, $+ 20$, $\div 2$, $- 1$, $\div 2$

h. Simplify: $z^2 - 1$ when $z = 1$

problem solving

A spinner has 4 sectors. If the spinner is spun once, what is the probability of the arrow stopping in sector *A*? sector *B*? sector *E*?

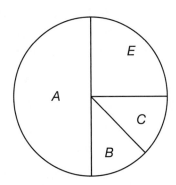

Recall that we find the area of a rectangle by multiplying its length by its width.

Area = length × width

This expression is a formula for finding the area of any rectangle. Usually formulas are written so that a letter represents each measure.

Below we list several common formulas. In these formulas, P stands for perimeter, and s represents the side length of a square.

Some Common Formulas

Area of a rectangle	$A = lw$
Perimeter of a rectangle	$P = 2(l + w)$ $P = 2l + 2w$
Area of a square	$A = s^2$
Perimeter of a square	$P = 4s$

Some figures are combinations of rectangles. In this example we see that the floor area of the house can be found by dividing the figure into rectangles and then adding the areas of the rectangles.

Example

Thinking Skills

Justify

Why can't we use the perimeter formula for this figure?

The diagram shows the blueprint of a one-story house.

a. What is the perimeter of the house?

b. What is the floor area of the house?

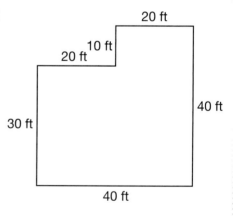

a. The perimeter of the house is the distance around the house. We add the lengths of the six sides.

30 + 40 + 40 + 20 + 10 + 20 = 160

Adding the lengths of the sides, we find that the perimeter of the house is **160 ft.**

b. To find the floor area we first divide the figure into two rectangles. We show one way to do this.

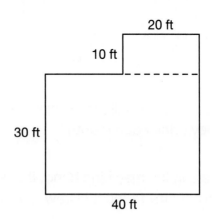

20 ft

10 ft

30 ft

40 ft

Thinking Skills

Verify

What formula did we use to find the area?

We have divided the figure with dashes, and we have labeled the length and width of both rectangles. Now we find the area of each rectangle.

$$
\begin{array}{r}
\text{Small rectangle} = 200\ \text{ft}^2 \\
+ \text{ Large rectangle} = 1200\ \text{ft}^2 \\
\hline
\text{Total Area of Figure} = 1400\ \text{ft}^2
\end{array}
$$

Adding the areas of the two rectangles, we find that the total floor area is **1400 sq. ft.**

Classify What is the name of the polygon in this example?

Lesson Practice Find the perimeter and area of each figure.

a.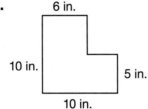

6 in.

10 in.

5 in.

10 in.

b.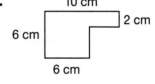

10 cm

2 cm

6 cm

6 cm

c.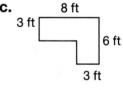

8 ft

3 ft

6 ft

3 ft

The figure at right shows the boundary of a garden. Refer to the figure to answer questions **d** and **e**.

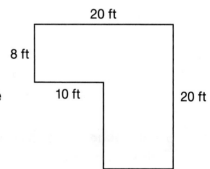

20 ft

8 ft

10 ft

20 ft

d. How many feet of wire fence are needed to enclose the garden along its boundary?

e. What is the area of the garden?

1. The Lorenzos drank 11 gallons of milk each month. How many quarts
(73) of milk did they drink each month?

2. Sixty people are in the marching band. If one fourth of them play trumpet,
(94) how many do not play trumpet? Draw a picture to illustrate the problem.

3. a. What is the area of this square?
(66)

b. What is the perimeter of the square?

10 mm

4. (**Analyze**) Esteban is 8 inches taller than Trevin. Trevin is 5 inches taller
(93) than Jan. Esteban is 61 inches tall. How many inches tall is Jan?

5. Which line segments in figure *ABCD* appear to be
(17, 33) parallel?

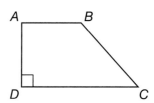

*6. Rename each fraction so that the denominator is 20.
(113)
    a. $\frac{1}{2}$ **b.** $\frac{3}{4}$ **c.** $\frac{3}{5}$

7. (**Predict**) If the arrow is spun, what is the probability that
(106, 110, it will stop on a number greater than 4?
Inv. 11)

*8. (**Estimate**) The asking price for the new house was $298,900. Round
(48) that amount of money to the nearest hundred thousand dollars.

9. (**Classify**) Name each of the following shapes. Then list the number of
(96, vertices, edges, and faces for each figure.
109)
    a.   **b.**

* **10.** Write the reduced form of each fraction:
(106)

  a. $\dfrac{9}{15}$ **b.** $\dfrac{10}{12}$ **c.** $\dfrac{12}{16}$

  11. (**Represent**) Use digits to write one hundred nineteen million, two
(104)   hundred forty-seven thousand, nine hundred eighty-four.

  12. $14.94 - (8.6 - 4.7)$ **13.** $6.8 - (1.37 + 2.2)$
(9, 45)   (9, 45)

* **14.** $3\dfrac{2}{5} + 1\dfrac{4}{5}$ * **15.** $\dfrac{5}{8} + \dfrac{3}{8}$ * **16.** $3\dfrac{2}{6} + 1\dfrac{1}{6}$
(112)   (98,   (112)
                  100)

* **17.** $5\dfrac{9}{10} - 1\dfrac{4}{10}$ * **18.** $\dfrac{5}{8} - \dfrac{3}{8}$ * **19.** $1 - \dfrac{1}{6}$
(112)   (112)   (98,
                                 100)

* **20.** 38×217 * **21.** 173×60 * **22.** 90×500
(107)   (107)   (85)

  23. $7\overline{)2942}$ **24.** $5\overline{)453}$ * **25.** $2\overline{)453}$
(83)   (75)   (79)

* **26.** (**Connect**) Segment AC is $3\dfrac{1}{2}$ inches long. Segment AB is $1\dfrac{1}{2}$ inches
(33, 100,
112) long. How long is segment BC?

A   B   C

* **27.** (**Estimate**) Fewer people live in Wyoming than in any other state.
(48)   According to the 2000 U.S. census, 493,782 people lived in
  Wyoming. Round this number of people to the nearest hundred
  thousand.

* **28.** (**Connect**) Rewrite $\dfrac{3}{4}$ and $\dfrac{5}{6}$ with a common denominator of 24.
(113)

* **29.** **Multiple Choice** If $s - 75 = t - 75$, which of the following is
(63)   true? Why?

  A $s > t$ **B** $s < t$ **C** $s = t$ **D** $s + 100 = t - 100$

*30. **(Analyze)** Workers are replacing a section of broken sidewalk. Before pouring the
(114)   concrete, the workers build a frame along the perimeter.

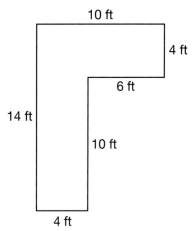

10 ft

4 ft

6 ft

14 ft

10 ft

4 ft

a. What is the perimeter of the replaced sidewalk?

b. What is the area of the replaced sidewalk?

A

acute angle
(17)

An angle whose measure is more than 0° and less than 90°.

acute angle   not **acute angles**

*An **acute angle** is smaller than both a right angle and an obtuse angle.*

ángulo agudo

Ángulo que mide más de 0° y menos de 90°.

*Un **ángulo agudo** es menor que un ángulo recto y que un ángulo obtuso.*

---

acute triangle
(81)

A triangle whose largest angle measures less than 90°.

acute triangle   not **acute triangles**

triángulo acutángulo

Triángulo cuyo ángulo mayor mide más que 0° y menos que 90°.

---

addend
(4)

Any one of the added numbers in an addition problem.

$2 + 3 = 5$ The **addends** in this problem are 2 and 3.

sumando

Cualquiera de los números en un problema de suma.

$2 + 3 = 5$ Los **sumandos** en este problema son el 2 y el 3.

---

addition
(12)

An operation that combines two or more numbers to find a total number.

$7 + 6 = 13$ We use **addition** to combine 7 and 6.

suma

Una operación que combina dos o mas números para encontrar un número total.

$7 + 6 = 13$ Usamos la **suma** para combinar el 7 y el 6.

---

a.m.
(13)

The period of time from midnight to just before noon.

*I get up at 7 **a.m.**, which is 7 o'clock in the morning.*

a.m.

Período de tiempo desde la medianoche hasta justo antes del mediodía.

*Me levanto a las 7 **a.m.**, lo cual es las 7 en punto de la mañana.*

---

angle
(17)

The opening that is formed when two lines, line segments, or rays intersect.

*These line segments form an **angle.***

ángulo

Abertura que se forma cuando se intersecan dos rectas, segmentos de recta o rayos.

*Estos segmentos de recta forman un **ángulo.***

apex *(96)*	The vertex (pointed end) of a cone or top of a pyramid.
ápice	El vértice (punta) de un cono.
approximation *(maintained)*	*See* **estimate.**
aproximación	*Ver* **estimar.**
area *(Inv. 3)*	The number of square units needed to cover a surface. 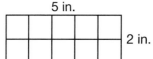 5 in. 2 in. *The **area** of this rectangle is 10 square inches.*
área	El número de unidades cuadradas que se necesita para cubrir una superficie. *El **área** de este rectángulo es de 10 pulgadas cuadradas.*
array *(Inv. 3)*	A rectangular arrangement of numbers or symbols in columns and rows. X X X X X X *This is a 3-by-4 **array** of X's.* X X X *It has 3 columns and 4 rows.* X X X
matriz	Un arreglo rectangular de números o símbolos en columnas y filas. *Esta es una **matriz** de X de 3 por 4. Tiene 3 columnas y 4 filas.*
Associative Property of Addition *(26)*	The grouping of addends does not affect their sum. In symbolic form, $a + (b + c) = (a + b) + c$. Unlike addition, subtraction is not associative. $(8 + 4) + 2 = 8 + (4 + 2)$ $(8 - 4) - 2 \neq 8 - (4 - 2)$ *Addition is **associative**.* *Subtraction is not **associative**.*
propiedad asociativa de la suma	La agrupación de los sumandos no altera la suma. En forma simbólica, $a + (b + c) = (a + b) + c$. A diferencia de la suma, la resta no es asociativa. $(8 + 4) + 2 = 8 + (4 + 2)$ $(8 - 4) - 2 \neq 8 - (4 - 2)$ *La suma es **asociativa**.* *La resta no es **asociativa**.*
Associative Property of Multiplication *(26)*	The grouping of factors does not affect their product. In symbolic form, $a \times (b \times c) = (a \times b) \times c$. Unlike multiplication, division is not associative. $(8 \times 4) \times 2 = 8 \times (4 \times 2)$ $(8 \div 4) \div 2 \neq 8 \div (4 \div 2)$ *Multiplication is **associative**.* *Division is not **associative**.*
propiedad asociativa de la multiplicación	La agrupación de los factores no altera el producto. En forma simbólica, $a \times (b \times c) = (a \times b) \times c$. A diferencia de la multiplicación, la división no es asociativa. $(8 \times 4) \times 2 = 8 \times (4 \times 2)$ $(8 \div 4) \div 2 \neq 8 \div (4 \div 2)$ *La multiplicación es **asociativa**.* *La división no es **asociativa**.*

average *(60)*	The number found when the sum of two or more numbers is divided by the number of addends in the sum; It is also called *mean*.

*To find the **average** of the numbers 5, 6, and 10, first add.*

$$5 + 6 + 10 = 21$$

Then, since there were three addends, divide the sum by 3.

$$21 \div 3 = 7$$

*The **average** of 5, 6, and 10 is 7.*

promedio	Número que se obtiene al dividir la suma de dos o más números entre la cantidad de sumandos; también se le llama *media*.

*Para calcular el **promedio** de los números 5, 6 y 10, primero se suman.*

$$5 + 6 + 10 = 21$$

Como hay tres sumandos, se divide la suma entre 3.

$$21 \div 3 = 7$$

*El **promedio** de 5, 6 y 10 es 7.*

B

bar graph *(Inv. 5)*	A graph that uses rectangles (bars) to show values or measurements.

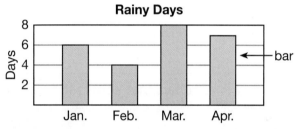

*This **bar graph** shows how many rainy days there were in each of these four months.*

gráfica de barras	Una gráfica que utiliza rectángulos (barras) para mostrar números o medidas.

*Esta **gráfica de barras** muestra cuántos días lluviosos hubo en cada uno de estos cuatro meses.*

base *(65)*	**1.** The lower number in an exponential expression.

$$\textbf{base} \longrightarrow 5^3 \longleftarrow exponent$$

5^3 *means* $5 \times 5 \times 5$*, and its value is 125.*

2. A designated side or face of a geometric figure.

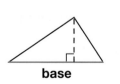

base   **base**   **base**

base	**1.** El número inferior en una expresión exponencial.

$$\textbf{base} \longrightarrow 5^3 \longleftarrow exponente$$

5^3 *significa* $5 \times 5 \times 5$*, y su valor es 125.*

2. Lado (o cara) determinado de una figura geométrica.

base-ten system *(2)*	A place-value system in which each place value is 10 times larger than the place value to its right. *The decimal system is a **base-ten system**.*
sistema base diez	Un sistema de valor posicional en el cual cada valor posicional es 10 veces mayor que el valor posicional que está a su derecha. *El sistema decimal es un **sistema base diez**.*
bias *(Inv. 6)*	Favoring one choice over another in a survey. *"Which do you prefer with lunch: cool, sweet lemonade or milk that has been out of the refrigerator for an hour?"* *Words like "cool" and "sweet" **bias** this survey question to favor the choice of lemonade.*
sesgo	Dar preferencia a una opción más que a otras en una encuesta. *"¿Qué prefieres tomar en tu almuerzo: una limonada dulce y fresca o leche que ha estado una hora fuera del refrigerador?" Palabras como "dulce" y "fresca" introducen **sesgo** en esta pregunta de encuesta para favorecer a la opción de limonada.*
borrowing *(maintained)*	*See **regrouping**.*
tomar prestado	*Ver **reagrupar**.*

C

calendar *(maintained)*	A chart that shows the days of the week and their dates.

SEPTEMBER 2007

S	M	T	W	T	F	S
						1
2	3	4	5	6	7	8
9	10	11	12	13	14	15
16	17	18	19	20	21	22
23	24	25	26	27	28	29
30						

calendar

calendario	Una tabla que muestra los días de la semana y sus fechas.
capacity *(73)*	The amount of liquid a container can hold. *Cups, gallons, and liters are units of **capacity**.*
capacidad	Cantidad de líquido que puede contener un recipiente. *Tazas, galones y litros son medidas de **capacidad**.*
cardinal numbers *(maintained)*	The counting numbers 1, 2, 3, 4, ...
números cardinales	Los números de conteo 1, 2, 3, 4, ...

Celsius
(21)

A scale used on some thermometers to measure temperature.

*On the **Celsius** scale, water freezes at 0°C and boils at 100°C.*

Celsius

Escala que se usa en algunos termómetros para medir la temperatura.

*En la escala **Celsius,** el agua se congela a 0°C y hierve a 100°C.*

---

center
(18)

The point inside a circle from which all points on the circle are equally distant.

*The **center** of circle A is 2 inches from every point on the circle.*

centro

Punto interior de un círculo o esfera, que equidista de cualquier punto del círculo o de la esfera.

*El **centro** del círculo A está a 2 pulgadas de cualquier punto del círculo.*

---

centimeter
(42)

One hundredth of a meter.

*The width of your little finger is about one **centimeter.***

centímetro

Una centésima de un metro.

*El ancho de tu dedo meñique mide aproximadamente un **centímetro.***

---

century
(maintained)

A period of one hundred years.

*The years 2001–2100 make up one **century.***

siglo

Un período de cien años.

*Los años 2001–2100 forman un **siglo.***

---

certain
(110)

We say that an event is *certain* when the event's probability is 1. This means the event will definitely occur.

seguro

Decimos que un suceso es *seguro* cuando la probabilidad del suceso es 1. Esto significa que el suceso ocurrirá definitivamente.

---

chance
(Inv. 11)

A way of expressing the likelihood of an event; It is the probability of an event expressed as a percent.

*The **chance** of rain is 20%. It is not likely to rain.*

*There is a 90% **chance** of snow. It is likely to snow.*

posibilidad

Modo de expresar la probabilidad de ocurrencia de un suceso; la probabilidad de un suceso expresada como porcentaje.

*La **posibilidad** de lluvia es del 20%. Es poco probable que llueva.*

*Hay un 90% de **posibilidad** de nieve. Es muy probable que nieve.*

chronological order *(maintained)*	The order of dates or times when listed from earliest to latest. *1951, 1962, 1969, 1973, 1981, 2001* *These years are listed in **chronological order.** They are listed from earliest to latest.*
orden cronológico	El orden de fechas o tiempos cuando se enlistan del más temprano al más tardío. *1952, 1962, 1969, 1973, 1981, 2001* *Estos años están listados en **orden cronológico.** Están listados del más temprano al más tardío.*
circle *(18)*	A closed, curved shape in which all points on the shape are the same distance from its center. 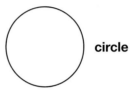 circle
círculo	Una forma cerrada curva en la cual todos los puntos en la figura están a la misma distancia de su centro.
circle graph *(Inv. 5)*	A graph made of a circle divided into sectors. Also called *pie chart* or *pie graph*. **Shoe Colors of Students** 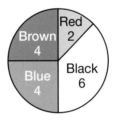 *This **circle graph** displays data on students' shoe color.*
gráfica circular	Una gráfica que consiste de un círculo dividido en sectores. También llamada *diagrama circular.* *Esta **gráfica circular** representa los datos de los colores de los zapatos de los estudiantes.*
circumference *(maintained)*	The distance around a circle; The perimeter of a circle. 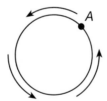 *If the distance from point A around to point A is 3 inches, then the **circumference** of the circle is 3 inches.*
circunferencia	La distancia alrededor de un círculo; el perímetro de un círculo. *Si la distancia desde el punto A alrededor del círculo hasta el punto A es 3 pulgadas, entonces la **circunferencia** del círculo mide 3 pulgadas.*

clockwise (78)	The same direction as the movement of a clock's hands.

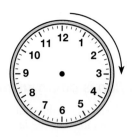 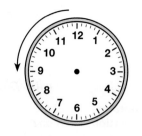

clockwise turn   counterclockwise turn

en el sentido de las manecillas del reloj	La misma dirección que el movimiento de las manecillas de un reloj.

combinations (36)	One or more parts selected from a set that are placed in groups in which order is not important.
	Combinations of the letters A, B, C, D, and E are AB, BC, CD, DE, AC, BD, CE, BE, and AE.
combinaciones	Una o mas partes seleccionadas de un conjunto que son colocadas en grupos donde el orden no es importante.

common denominators (113)	Denominators that are the same.
	The fractions $\frac{2}{5}$ and $\frac{3}{5}$ have **common denominators.**
denominadores comunes	Denominadores que son iguales.
	Las fracciones $\frac{2}{5}$ y $\frac{3}{5}$ tienen **denominadores comunes.**

common year (maintained)	A year with 365 days; Not a leap year.
	The year 2000 is a leap year, but 2001 is a **common year.** In a **common year** February has 28 days. In a leap year it has 29 days.
año común	Un año con 365 días; no un año bisiesto.
	El año 2000 es un año bisiesto, pero 2001 es un **año común.** En un **año común** febrero tiene 28 días. En un año bisiesto tiene 29 días.

Commutative Property of Addition (maintained)	Changing the order of addends does not change their sum. In symbolic form, $a + b = b + a$. Unlike addition, subtraction is not commutative.
	$8 + 2 = 2 + 8$ $8 - 2 \neq 2 - 8$
	Addition is **commutative.** Subtraction is not **commutative.**
propiedad conmutativa de la suma	El orden de los sumandos no altera la suma. En forma simbólica, $a + b = b + a$. A diferencia de la suma, la resta no es conmutativa.
	$8 + 2 = 2 + 8$ $8 - 2 \neq 2 - 8$
	La suma es **conmutativa.** La resta no es **conmutativa.**

Commutative Property of Multiplication (23)	Changing the order of factors does not change their product. In symbolic form, $a \times b = b \times a$. Unlike multiplication, division is not *commutative*.
	$$8 \times 2 = 2 \times 8 \qquad\qquad 8 \div 2 \neq 2 \div 8$$
	*Multiplication is **commutative**.* *Division is not **commutative**.*
propiedad conmutativa de la multiplicación	El orden de los factores no altera el producto. En forma simbólica, $a \times b = b \times a$. A diferencia de la multiplicación, la división no es *conmutativa*.
	$$8 \times 2 = 2 \times 8 \qquad\qquad 8 \div 2 \neq 2 \div 8$$
	*La multiplicación es **conmutativa**.* *La división no es **conmutativa**.*

comparing (Inv. 1)	To determine if two numbers are equal, or if one number is greater than or less than another number.
comparar	Determinar si dos numeros son iguales, o si uno es mayor o menor que el otro.

comparison symbol (Inv. 2)	A mathematical symbol used to compare numbers.
	***Comparison symbols** include the equal sign (=) and the "greater than/less than" symbols ($>$ or $<$).*
símbolo de comparación	Un símbolo matemático que se usa para comparar números.
	Los símbolos de comparación incluyen el signo de igualdad (=) y los símbolos de "mayor que/menor que" ($>$ ó $<$).

compass (18)	A tool used to draw circles and arcs.

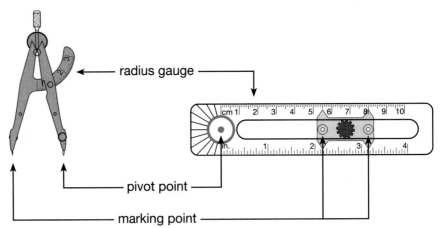

two types of **compasses**

compás	Instrumento para dibujar círculos y arcos.

compatible numbers (37)	Numbers that are close in value to the actual numbers and are easy to add, subtract, multiply, or divide.
números compatibles	Números que tienen un valor cercano a los números reales y que son fáciles de sumar, restar, multiplicar, o dividir.

composite numbers *(56)*	A counting number greater than 1 that is divisible by a number other than itself and 1. Every *composite number* has three or more factors. Every *composite number* can be expressed as a product of two or more prime numbers. *9 is divisible by 1, 3, and 9. It is **composite.*** *11 is divisible by 1 and 11. It is not **composite.***
números compuestos	Un número de conteo mayor que 1, divisible entre algún otro número distinto de sí mismo y de 1. Cada *número compuesto* tiene tres o más factores. Cada *número compuesto* puede ser expresado como el producto de dos o más números primos. *9 es divisible entre 1, 3 y 9. Es **compuesto.*** *11 es divisible entre 1 y 11. No es **compuesto.***

cone *(96)*	A three-dimensional solid with one curved surface and one flat, circular surface. The pointed end of a *cone* is its apex.

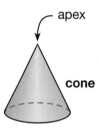

apex
cone

cono	Un sólido tridimensional con una superficie curva y una superficie plana y circular. El extremo puntiagudo de un *cono* es su ápice.

congruent *(57)*	Having the same size and shape.

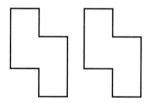

*These polygons are **congruent.** They have the same size and shape.*

congruentes	Que tienen igual tamaño y forma. *Estos polígonos son **congruentes.** Tienen igual tamaño y forma.*

continuous *(Inv. 5)*	Data that can be any value, such as measures of distance, time, or weight. *A line graph is appropriate for displaying **continuous** data.*
continuos	Datos que pueden tener cualquier valor, como medidas de distancia, tiempo o de peso. *Una gráfica linal es apropiada para mostrar datos **continuos.***

coordinate(s) *(Inv. 7)*	**1.** A number used to locate a point on a number line.

A

-3 -2 -1 0 1 2 3

*The **coordinate** of point A is −2.*

2. A pair of numbers used to locate a point on a *coordinate* plane.

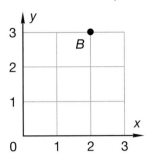

*The **coordinates** of point B are (2, 3). The **x-coordinate** is listed first, and the **y-coordinate** is listed second.*

coordenada(s)

1. Número que se utiliza para ubicar un punto sobre una recta numérica.

La **coordenada** del punto A es −2.

2. Par ordenado de números que se utiliza para ubicar un punto sobre un plano coordenado.

Las **coordenadas** del punto B son (2, 3). La **coordenada x** se escribe primero, seguida de la **coordenada y**.

---

coordinate plane
(Inv. 7)

A grid on which any point can be identified by its distances from the *x*- and *y*-axes.

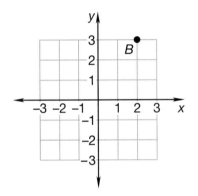

plano coordenado

Una cuadrícula en la cual cualquier punto puede ser identificado por sus distancias de los ejes x y y.

---

counter-clockwise
(78)

The direction opposite of the movement of a clock's hands.

counterclockwise turn   clockwise turn

en sentido contrario a las manecillas del reloj

La dirección opuesta al movimiento de las manecillas de un reloj.

counting numbers *(1)*	The numbers used to count; The numbers in this sequence: 1, 2, 3, 4, 5, 6, 7, 8, 9, ... *The numbers 12 and 37 are* **counting numbers,** *but 0.98 and $\frac{1}{2}$ are not.*
números de conteo	Números que se utilizan para contar; los números en esta secuencia: 1, 2, 3, 4, 5, 6, 7, 8, 9, ... *Los números 12 y 37 son* **números de conteo,** *pero 0.98 y $\frac{1}{2}$ no son* **números de conteo.**
cube *(96)*	A three-dimensional solid with six square faces. Adjacent faces are perpendicular and opposite faces are parallel.

cube

cubo	Un sólido tridimensional con seis caras cuadradas. Las caras adyacentes son perpendiculares y las caras opuestas son paralelas.
cylinder *(96)*	A three-dimensional solid with two circular bases that are opposite and parallel to each other.

cylinder

cilindro	Un sólido tridimensional con dos bases circulares que son opuestas y paralelas entre sí.

D

data *(Inv. 5)*	(Singular: *datum*) Information gathered from observations or calculations. *82, 76, 95, 86, 98, 97, 93* *These* **data** *are average daily temperatures for one week in Utah.*
datos	Información reunida de observaciones o cálculos. *Estos* **datos** *son el promedio diario de las temperaturas de una semana en Utah.*
decade *(maintained)*	A period of ten years. *The years 2001–2010 make up one* **decade.**
década	Un periodo de diez años. *Los años 2001–2010 forman una* **década.**

decagon *(50)*	A polygon with ten sides.

decagon

decágono	Un polígono de diez lados.

decimal number *(Inv. 4)*	A numeral that contains a decimal point. *23.94 is a **decimal number** because it contains a decimal point.*
número decimal	Número que contiene un punto decimal. *23.94 es un **número decimal**, porque tiene punto decimal.*

decimal place(s) *(Inv. 4)*	Places to the right of a decimal point. *5.47 has two **decimal places.*** *6.3 has one **decimal place.*** *8 has no **decimal places.***
cifras decimales	Lugares ubicados a la derecha del punto decimal. *5.47 tiene dos **cifras decimales.*** *6.3 tiene una **cifra decimal.*** *8 no tiene **cifras decimales.***

decimal point *(28)*	A symbol used to separate the ones place from the tenths place in decimal numbers (or dollars from cents in money).

$$34.15$$

↑

decimal point

punto decimal	Un símbolo que se usa para separar el lugar de las unidades del lugar de la decenas en números decimales (o los dólares de los centavos en dinero).

degree (°)
(Inv. 2, 78)

1. A unit for measuring temperature.

Water boils.

*There are 100 **degrees** (100°) between the freezing and boiling points of water on the Celsius scale.*

Water freezes.

2. A unit for measuring angles.

90°

*There are 90 **degrees** (90°) in a right angle.*

grado (°)

1. Unidad para medir la temperatura.

*Hay 100 **grados** de diferencia entre los puntos de ebullición y congelación del agua en la escala Celsius, o escala centígrada.*

2. Unidad para medir ángulos.

*Un ángulo recto mide 90 **grados** (90°).*

---

denominator
(19)

The bottom number of a fraction; It is the number that tells how many parts are in a whole.

$\frac{1}{4}$

*The **denominator** of the fraction is 4. There are 4 parts in the whole circle.*

denominador

El número inferior de una fracción; el número que indica cuántas partes hay en un entero.

*El **denominador** de la fracción es 4. Hay 4 partes en el círculo completo.*

---

diameter
(18)

The distance across a circle through its center.

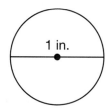

1 in.

*The **diameter** of this circle is 1 inch.*

diámetro

Distancia que atravieza un círculo a través de su centro.

*El **diámetro** de este círculo mide 1 pulgada.*

difference *(maintained)*	The result of subtraction. $12 - 8 = 4$ The **difference** in this problem is 4.
diferencia	Resultado de una resta. $12 - 8 = 4$ La **diferencia** en este problema es 4.

---

digit *(1)*	Any of the symbols used to write numbers: 0, 1, 2, 3, 4, 5, 6, 7, 8, 9. *The last **digit** in the number 2587 is 7.*
dígito	Cualquiera de los símbolos que se utilizan para escribir números: 0, 1, 2, 3, 4, 5, 6, 7, 8, 9. *El último **dígito** del número 2587 es 7.*

---

digital form *(13)*	When referring to clock time, *digital form* is a way to write time that uses a colon and a.m. or p.m. *11:30 a.m. is **digital form.***
forma digital	Cuando nos referimos al tiempo marcado por un reloj, la *forma digital* es una manera de escribir tiempo que usa dos puntos y a.m. o p.m. *11:30 a.m. está en **forma digital.***

Distributive Property *(102)*	A number times the sum of two addends is equal to the sum of that same number times each individual addend. $a \times (b + c) = (a \times b) + (a \times c)$ $8 \times (2 + 3) = (8 \times 2) + (8 \times 3)$ $8 \times 5 = 16 + 24$ $40 = 40$ *Multiplication is **distributive** over addition.*
propiedad distributiva	Un número multiplicado por la suma de dos sumandos es igual a la suma de los productos de ese número por cada uno de los sumandos. $a \times (b + c) = (a \times b) + (a \times c)$ $8 \times (2 + 3) = (8 \times 2) + (8 \times 3)$ $8 \times 5 = 16 + 24$ $40 = 40$ *La multiplicación es **distributiva** con respecto a la suma.*

---

dividend *(54)*	A number that is divided. $12 \div 3 = 4$ $3\overline{)12}\,^{4}$ $\dfrac{12}{3} = 4$ *The **dividend** is 12 in each of these problems.*
dividendo	Número que se divide. $12 \div 3 = 4$ $3\overline{)12}\,^{4}$ $\dfrac{12}{3} = 4$ *El **dividendo** es 12 en cada una de estas operaciones.*

divisible
(55)

Able to be divided by a whole number without a remainder.

$$4\overline{)20} \;\; 5$$

*The number 20 is **divisible** by 4, since 20 ÷ 4 has no remainder.*

$$3\overline{)20} \;\; 6\,R\,2$$

*The number 20 is not **divisible** by 3, since 20 ÷ 3 has a remainder.*

divisible

Número que se puede dividir exactamente por un entero, es decir, sin residuo.

$$4\overline{)20} \;\; 5$$

*El número 20 es **divisible** entre 4, ya que 20 ÷ 4 no tiene residuo.*

$$3\overline{)20} \;\; 6\,R\,2$$

*El número 20 no es **divisible** entre 3, ya que 20 ÷ 3 tiene residuo.*

division
(34)

An operation that separates a number into a given number of equal parts or into a number of parts of a given size.

$$21 \div 3 = 7$$

*We use **division** to separate 21 into 3 groups of 7.*

división

Una operación que separa un número en un número dado de partes iguales o en un número de partes de una medida dada.

*Usamos la **división** para separar 21 en 3 grupos de 7.*

divisor
(54)

A number by which another number is divided.

$$12 \div 3 = 4 \qquad 3\overline{)12} \;\; 4 \qquad \frac{12}{3} = 4$$

*The **divisor** is 3 in each of these problems.*

divisor

Número que divide a otro en una división.

$$12 \div 3 = 4 \qquad 3\overline{)12} \;\; 4 \qquad \frac{12}{3} = 4$$

*El **divisor** es 3 en cada una de estas operaciones.*

dozen
(39)

A group of twelve.

*The carton holds a **dozen** eggs.*

The carton holds 12 eggs.

docena

Un grupo de doce.

*El cartón contiene una **docena** de huevos.*

El cartón contiene 12 huevos.

E

edge
(96)

A line segment formed where two faces of a solid intersect.

*The arrow is pointing to one **edge** of this cube. A cube has 12 **edges.***

arista

Segmento de recta formado donde se intersecan dos caras de un sólido.

*La flecha apunta hacia una **arista** de este cubo. Un cubo tiene 12 **aristas.***

elapsed time *(13)*	The difference between a starting time and an ending time. *The race started at 6:30 p.m. and finished at 9:12 p.m. The **elapsed time** of the race was 2 hours 42 minutes.*
tiempo transcurrido	La diferencia entre el tiempo de comienzo y tiempo final. *La carrera comenzó a las 6:30 p.m. y terminó a las 9:12 p.m . El **tiempo transcurrido** de la carrera fue de 2 horas 42 minutos.*
elevation *(9)*	A measure of distance above sea level. *At it's peak, Mount Whitney in California has an **elevation** of 14,494 ft above sea level.*
elevación	Una medida de distancia arriba del nivel del mar. *En su cima, Mount Whitney en California está 14,494 pies arriba del nivel del mar.*
endpoint(s) *(17)*	The point(s) at which a line segment ends. $$A \bullet\!\!\!-\!\!\!-\!\!\!-\!\!\!-\!\!\!-\!\!\!-\!\!\!-\!\!\!-\!\!\!-\!\!\!\bullet\, B$$ *Points A and B are the **endpoints** of line segment AB.*
punto(s) extremo(s)	Punto(s) donde termina un segmento de recta. *Los puntos A y B son los **puntos extremos** del segmento AB.*
equal to *(Inv. 2)*	Has the same value as. *12 inches are **equal to** 1 foot.*
es igual a	Con el mismo valor. *12 pulgadas **es igual a** 1 pie.*
equation *(4)*	A number sentence that uses an equal sign (=) to show that two quantities are equal. $x = 3 \qquad 3 + 7 = 10 \qquad\qquad 4 + 1 \qquad x < 7$ **equations** not **equations**
ecuación	Enunciado que usa el símbolo "=" para indicar que dos cantidades son iguales. $x = 3 \qquad 3 + 7 = 10 \qquad\qquad 4 + 1 \qquad x < 7$ son **ecuaciones** no son **ecuaciones**
equiangular *(81)*	A figure with angles of the same measurement. *An equilateral triangle is also **equiangular** because its angles each measure 60°.*
equiangular	Una figura con ángulos de la misma medida. *Un triángulo equilátero es también **equiangular** porque sus tres ángulos miden 60°.*

equilateral triangle *(18)*	A triangle in which all sides are the same length and all angles are the same measure.

*This is an **equilateral triangle.***
All of its sides are the same length.
All of its angles are the same measure.

triángulo equilátero	Triángulo que tiene todos sus lados de la misma longitud.

*Éste es un **triángulo equilátero.** Sus tres lados tienen la misma longitud. Todos sus ángulos miden los mismo.*

---

equivalent fractions *(103)*	Different fractions that name the same amount.

$\frac{1}{2}$  = 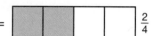 $\frac{2}{4}$

$\frac{1}{2}$ *and* $\frac{2}{4}$ *are **equivalent fractions.***

fracciones equivalentes	Fracciones diferentes que representan la misma cantidad.

$\frac{1}{2}$ *y* $\frac{2}{4}$ *son **fracciones equivalentes.***

---

estimate *(22)*	To find an approximate value.

*I **estimate** that the sum of 203 and 304 is about 500.*

estimar	Encontrar un valor aproximado.

*Puedo **estimar** que la suma de 199 más 205 es aproximadamente 400.*

---

evaluate *(9)*	To find the value of an expression.

*To **evaluate** a + b for a = 7 and b = 13, we replace a with 7 and b with 13:*

$$7 + 13 = 20$$

evaluar	Calcular el valor de una expresión.

*Para **evaluar** a + b, con a = 7 y b = 13, se reemplaza a por 7 y b por 13:*
$$7 + 13 = 20$$

---

even numbers *(1)*	Numbers that can be divided by 2 without a remainder; the numbers in this sequence: 0, 2, 4, 6, 8, 10, ...

***Even numbers** have 0, 2, 4, 6, or 8 in the ones place.*

números pares	Números que se pueden dividir entre 2 sin residuo; los números en esta secuencia: 0, 2, 4, 6, 8, 10, ...

*Los **números pares** terminan en 0, 2, 4, 6, u 8 en el lugar de las unidades.*

---

exchanging *(maintained)*	*See **regrouping.***
cambiar	*Ver **reagrupar.***

---

expanded form *(5)*	A way of writing a number that shows the value of each digit.

*The **expanded form** of 234 is 200 + 30 + 4.*

forma desarrollada	Una manera de escribir un número mostrando el valor de cada dígito.

*La **forma desarrollada** de 234 es 200 + 30 + 4.*

exponent *(65)*	The upper number in an exponential expression; it shows how many times the base is to be used as a factor. $$\text{base} \longrightarrow 5^3 \longleftarrow \textbf{\textit{exponent}}$$ *5^3 means $5 \times 5 \times 5$, and its value is 125.*
exponente	El número superior en una expresión exponencial; muestra cuántas veces debe usarse la base como factor. $$\text{base} \longrightarrow 5^3 \longleftarrow \textit{exponente}$$ *5^3 significa $5 \times 5 \times 5$, y su valor es 125.*
exponential expression *(65)*	An expression that indicates that the base is to be used as a factor the number of times shown by the exponent. $$4^3 = 4 \times 4 \times 4 = 64$$ *The **exponential expression** 4^3 uses 4 as a factor 3 times. Its value is 64.*
expresión exponencial	Expresión que indica que la base debe usarse como factor el número de veces que indica el exponente. $$4^3 = 4 \times 4 \times 4 = 64$$ *La **expresión exponencial** 4^3 se calcula usando 3 veces el 4 como factor. Su valor es 64.*
expression *(9)*	A number, a letter, or a combination of both. *Expressions* do not include comparison symbols, such as an equal sign. *3n is an **expression** that can also be written as $3 \times n$.*
expresión	Un número, una letra o una combinación de los dos. Las *expresiones* no incluyen símbolos de comparación, como el signo de igual. *3n es una **expresión** que también puede ser escrita como $3 \times n$.*

F

face *(96)*	A flat surface of a geometric solid. *The arrow is pointing to one **face** of the cube. A cube has six **faces**.*
cara	Superficie plana de un cuerpo geométrico. *La flecha apunta a una **cara** del cubo. Un cubo tiene seis **caras**.*
fact family *(maintained)*	A group of three numbers related by addition and subtraction or by multiplication and division. *The numbers 3, 4, and 7 are a **fact family**. They make these four facts:* $$3 + 4 = 7 \qquad 4 + 3 = 7 \qquad 7 - 3 = 4 \qquad 7 - 4 = 3$$

familia de operaciones
Grupo de tres números relacionados por sumas y restas o por multiplicaciones y divisiones.

*Los números 3, 4 y 7 forman una **familia de operaciones**. Con ellos se pueden formar estas cuatro operaciones:*

$$3 + 4 = 7 \qquad 4 + 3 = 7 \qquad 7 - 3 = 4 \qquad 7 - 4 = 3$$

factor
(23)
Any one of the numbers multiplied in a multiplication problem.

$2 \times 3 = 6$ *The **factors** in this problem are 2 and 3.*

factor
Cualquier número que se multiplica en un problema de multiplicación.

$2 \times 3 = 6$ *Los **factores** en este problema son 2 y 3.*

Fahrenheit
(21)
A scale used on some thermometers to measure temperature.

*On the **Fahrenheit** scale, water freezes at 32°F and boils at 212°F.*

Fahrenheit
Escala que se usa en algunos termómetros para medir la temperatura.

*En la **escala Fahrenheit**, el agua se congela a 32°F y hierve a 212°F.*

fluid ounce
(73)
A unit of liquid measurement in the customary system.

*There are 8 **fluid ounces** in a cup, 16 **fluid ounces** in a pint, and 32 **fluid ounces** in a quart.*

onza líquida (oz. liq.)
Una unidad de medida para líquidos en el sistema usual.

*Hay 8 **onzas líquidas** en una taza, 16 **onzas líquidas** en una pinta y 32 **onzas líquidas** en un cuarto.*

formula
(12)
An expression or equation that describes a method for solving a certain type of problem. We often write *formulas* with letters that stand for complete words.

*A **formula** for the perimeter of a rectangle is P = 2l + 2w, where P stands for "perimeter," l stands for "length," and w stands for "width."*

fórmula
Una expresión o ecuación que describe un método para resolver cierto tipo de problemas. Frecuentemente escribimos *fórmulas* con letras que representan palabras completas.

*Una **fórmula** para el perímetro de un rectángulo es P = 2l + 2w, donde P representa "perímetro", l representa "longitud" y w representa "ancho".*

fraction
(19)
A number that names part of a whole.

$\frac{1}{4}$ *of the circle is shaded.*

$\frac{1}{4}$ *is a **fraction**.*

fracción
Número que representa una parte de un entero.

$\frac{1}{4}$ *del círculo está sombreado.*

$\frac{1}{4}$ *es una **fracción**.*

full turn
(78)
A turn measuring 360°.

giro completo
Giro que mide 360°.

G

geometric solid
(96)

A shape that takes up space.

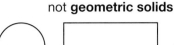

geometric solids
cube   cylinder

not **geometric solids**
circle   rectangle   hexagon

sólido geométrico

Una figura que ocupa espacio.

---

geometry
(maintained)

A major branch of mathematics that deals with shapes, sizes, and other properties of figures.

*Some of the figures we study in **geometry** are angles, circles, and polygons.*

geometría

Rama extensa de las matemáticas que trata de las formas, tamaños y otras propiedades de las figuras.

*Algunas de las figuras que se estudian en **geometría** son los ángulos, círculos y polígonos.*

---

graph
(Inv. 5)

A diagram that shows data in an organized way. *See also* **bar graph, circle graph, line graph,** *and* **pictograph.**

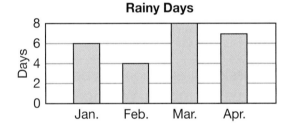

bar **graph**

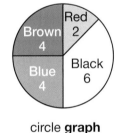

circle **graph**

gráfica

Diagrama que muestra datos de una forma organizada. *Ver también **Gráfica de barras, gráfica circular, gráfica lineal, y pictograma.***

---

greater than
(Inv. 2)

Having a larger value than.

$5 > 3$ *Five is **greater than** three.*

mayor que

Que tiene un valor mayor que.

$5 > 3$ *Cinco es **mayor que** tres.*

H

half
(19)

One of two equal parts that together equal a whole.

mitad

Una de dos partes iguales que juntas forman un todo.

---

half turn
(78)

A turn measuring 180°.

medio giro

Un giro que mide 180°.

hexagon *(20)*	A polygon with six sides.

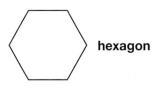

hexágono	Un polígono con seis lados.

---

horizontal *(Inv. 7)*	Side to side; perpendicular to vertical

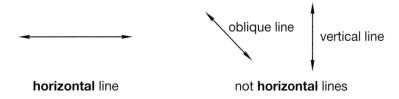

horizontal	Lado a lado; perpendicular a la vertical

---

hundredth(s) *(Inv. 4)*	One of one hundred parts. *The decimal form of one **hundredth** is 0.01.*
centésima(s)	Una de cien partes. *La forma decimal de una **centésima** es 0.01.*

I

Identity Property **of Addition** *(maintained)*	The sum of any number and 0 is equal to the initial number. In symbolic form, $a + 0 = a$. The number 0 is referred to as the *additive identity.* *The **Identity Property of Addition** is shown by this statement:* $$13 + 0 = 13$$
propiedad de identidad **de la suma**	La suma de cualquier número más 0 es igual al número inicial. En forma simbólica, $a + 0 = a$. El número 0 se conoce como *identidad aditiva.* *La **propiedad de identidad de la suma** se muestra en el siguiente enunciado:* $$13 + 0 = 13$$

Identity Property **of Multiplication** *(23)*	The product of any number and 1 is equal to the initial number. In symbolic form, $a \times 1 = a$. The number 1 is referred to as the *multiplicative identity.* *The **Identity Property of Multiplication** is shown by this statement:* $$94 \times 1 = 94$$
propiedad de identidad **de la multiplicación**	El producto de cualquier número por 1 es igual al número inicial. En forma simbólica, $a \times 1 = a$. El número 1 se conoce como *identidad multiplicativa.* *La **propiedad de identidad de la multiplicación** se muestra en el siguiente enunciado:* $$94 \times 1 = 94$$

improper fraction *(89)*	A fraction with a numerator greater than or equal to the denominator.
	$\dfrac{4}{3}$ $\dfrac{2}{2}$ *These fractions are **improper fractions.***
fracción impropia	Fracción con el numerador igual o mayor que el denominador.
	$\dfrac{4}{3}$ $\dfrac{2}{2}$ *Estas fracciones son **fracciones impropias.***

intersect *(17)*	To share a common point or points.
	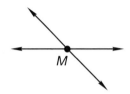 *These two lines **intersect.** They share the common point M.*
intersecar	Compartir uno o varios puntos en común.
	*Estas dos rectas se **intersecan.***
	Tienen el punto común M.

intersecting lines *(17)*	Lines that cross.
	intersecting lines
líneas que se cruzan o intersecan	Líneas que se cruzan.

inverse operation(s) *(8)*	An operation that "undoes" another.
	*Subtraction is the **inverse operation** of addition.*
operaciones inversas	Una operación que cancela a otra.
	*La resta es la **operación inversa** de la suma.*

isosceles triangle *(81)*	A triangle with at least two sides of equal length and two angles of equal measure.
	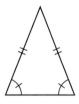 *Two of the sides of this **isosceles triangle** have equal lengths. Two of the angles have equal measures.*
triángulo isósceles	Triángulo que tiene por lo menos dos lados de igual longitud y dos lados de igual medida.
	*Dos de los lados de este **triángulo isósceles** tienen igual longitud.*
	Dos de los ángulos tienen medidas iguales.

K

key
(Inv.5)

See **legend.**

clave

Ver **rótulo.**

kilometer
(42)

A metric unit of length equal to 1000 meters.

*One **kilometer** is approximately 0.62 miles.*

kilómetro

Una unidad métrica de longitud igual a 1000 metros.

Un kilómetro es aproximadamente 0.62 milla.

L

leap year
(maintained)

A year with 366 days; not a common year

*In a **leap year,** February has 29 days.*

año bisiesto

Un año con 366 dias; no un año común

*En un **año bisiesto** febrero tiene 29 días.*

legend
(Inv. 5)

A notation on a map, graph, or diagram that describes the meaning of the symbols and/or the scale used.

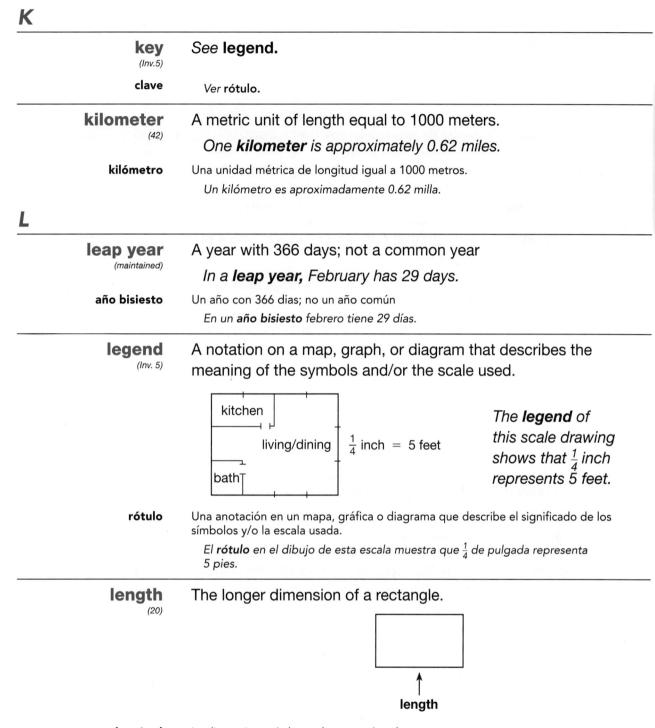

kitchen

living/dining $\frac{1}{4}$ inch = 5 feet

bath

*The **legend** of this scale drawing shows that $\frac{1}{4}$ inch represents 5 feet.*

rótulo

Una anotación en un mapa, gráfica o diagrama que describe el significado de los símbolos y/o la escala usada.

*El **rótulo** en el dibujo de esta escala muestra que $\frac{1}{4}$ de pulgada representa 5 pies.*

length
(20)

The longer dimension of a rectangle.

↑
length

longitud

La dimension más larga de un rectángulo.

less than
(Inv. 2)

Having a smaller value than.

$3 < 5$ *Three is **less than** five.*

menor que

Con un valor menor que.

$3 < 5$ *Tres es **menor que** cinco.*

line (17)	A straight collection of points extending in opposite directions without end. 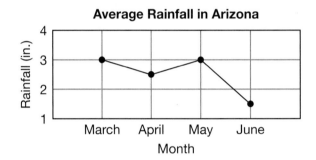 **line** *AB* or **line** *BA*
recta	Una grupo de puntos en línea recta que se extienden sin fin en direcciones opuestas.

linear unit (31)	Units used to measure lengths and distances.
unidad linear	Unidades que se usan para medir longitudes y distancias.

line graph (Inv. 5)	A graph that connects points to show how information changes over time. **Average Rainfall in Arizona** 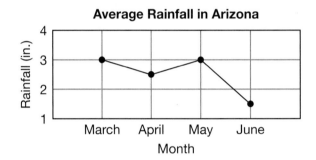 *This **line graph** shows the average rainfall in Arizona over four months.*
gráfica lineal	Una gráfica que conecta puntos para mostrar como la información cambia con el tiempo. *Esta **gráfica lineal** muestra el promedio de lluvias en Arizona en un periodo de cuatro meses.*

line of symmetry (82)	A line that divides a figure into two halves that are mirror images of each other. *See also* **symmetry.** **lines of symmetry** not **lines of symmetry**
eje de simetría	Una línea que divide una figura en dos mitades que son imágenes especulares una de otra. *Ver también **simetría.***

line segment (17)	A part of a line with two distinct endpoints.  $\overline{AB}$ *is a **line segment.***
segmento de recta	Una parte de una línea con dos extremos específicos. $\overline{AB}$ *es un **segmento de recta.***

liter (73)	A metric unit of capacity or volume. *A **liter** is a little more than a quart.*
litro	Una unidad métrica de capacidad o volumen. *Un **litro** es un poco más que un cuarto.*
lowest terms (106)	A fraction is in *lowest terms* if it cannot be reduced. *In **lowest terms,** the fraction $\frac{8}{20}$ is $\frac{2}{5}$.*
mínima expresión	Una fracción está en su mínima expresión si no se puede reducir. *En su **mínima expresión** la fracción $\frac{8}{20}$ es $\frac{2}{5}$.*

M

mass (80)	The amount of matter an object contains. A kilogram is a metric unit of *mass*. *The **mass** of a bowling ball would be the same on the moon as on Earth, even though the weight of the bowling ball would be different.*
masa	La cantidad de materia que contiene un objeto. Un kilogramo es una unidad métrica de *masa*. *La **masa** de una bola de boliche sería la misma en la Luna que en la Tierra. Aunque el peso de la bola de boliche sería diferente.*
median (Inv. 6)	The middle number (or the average of the two central numbers) of a list of data when the numbers are arranged in order from least to the greatest. *1, 1, 2, 4, 5, 7, 9, 15, 24, 36, 44* *In this list of data, 7 is the **median.***
mediana	Número de en medio (o el promedio de los dos números centrales) en una lista de datos, cuando los números se ordenan de menor a mayor. *1, 1, 2, 4, 5, 7, 9, 15, 24, 36, 44* *En esta lista de datos, 7 es la **mediana.***
meter (42)	The basic unit of length in the metric system. *A **meter** is equal to 100 centimeters, and it is slightly longer than 1 yard.* *Many classrooms are about 10 **meters** long and 10 **meters** wide.*
metro	La unidad básica de longitud en el sistema métrico *Un **metro** es igual a 100 centímetros y es un poco más largo que una yarda.* *Muchos salones de clase son de alrededor de 10 **metros** de largo y 10 **metros** de ancho.*

metric system *(42)*	An international system of measurement in which units are related by a power of ten. Its also called the *International System of Measurement.* *Centimeters and kilograms are units in the **metric system.***
sistema métrico	Un sistema internacional de medidas en donde las unidades se relacionan con una potencia de diez. También llamado el *Sistema internacional.* *Los centímetros y los kilogramos son unidades del **sistema métrico.***
midnight *(13)*	12:00 a.m. ***Midnight** is one hour after 11 p.m.*
medianoche	12:00 a.m. *La **medianoche** es una hora después de las 11 p.m.*
millimeter *(42)*	A metric unit of length. *There are 1000 **millimeters** in 1 meter and 10 **millimeters** in 1 centimeter.*
milímetro	Una unidad métrica de longitud. *Hay 1000 **milímetros** en 1 metro y 10 **milímetros** en 1 centímetro.*
mixed number *(32)*	A number expressed as a whole number plus a fraction. *The **mixed number** $5\frac{3}{4}$ means "five and three fourths."*
número mixto	Un número expresado como un número entero más una fracción. *El **número mixto** $5\frac{3}{4}$ significa "cinco y tres cuartos."*
mode *(Inv. 6)*	The number or numbers that appear most often in a list of data. *5, 12, 32, 5, 16, 5, 7, 12* *In this list of data, the number 5 is the **mode.***
moda	Número o números que aparecen con más frecuencia en una lista de datos. *5, 12, 32, 5, 16, 5, 7, 12* *En esta lista de datos, el número 5 es la **moda.***
multiple *(22)*	A product of a counting number and another number. *The **multiples** of 3 include 3, 6, 9, and 12.*
múltiplo	Producto de un número de conteo y otro número *Los **múltiplos** de 3 incluyen 3, 6, 9 y 12.*
multiplication *(23)*	An operation that uses a number as an addend a specified number of times. $7 \times 3 = 21$ *We can use **multiplication** to* $7 + 7 + 7 = 21$ *use 7 as an addend 3 times.*
multiplicación	Una operación que usa un número como sumando un número específico de veces. $7 \times 3 = 21$ *Podemos usar la **multiplicación** para usar* $7 + 7 + 7 = 21$ *el 7 como sumando 3 veces.*

multiplication table *(23)*	A table used to find the product of two numbers. The product of two numbers is found at the intersection of the row and the column for the two numbers.
tabla de multiplicación	Una tabla que se usa para encontrar el producto de dos números. El producto de dos números se encuentra en la intersección de la fila y la columna para los dos números.

N

negative numbers *(Inv. 2)*	Numbers less than zero. *−15 and −2.86 are **negative numbers.*** *19 and 0.74 are not **negative numbers.***
números negativos	Los números menores que cero. *−15 y −2.86 son **números negativos.*** *19 y 0.74 no son **números negativos.***

net *(97)*	An arrangement of edge-joined polygons that can be folded to become the faces of the geometric solid. 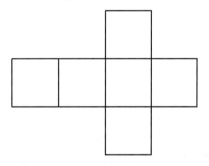
red	Un arreglo de polígonos unidos por el borde que pueden ser doblados para convertirse en las caras de un sólido geométrico.

noon *(13)*	12:00 p.m. ***Noon** is one hour after 11 a.m.*
mediodía	12:00 p.m. ***Mediodía** es una hora después de las 11 a.m.*

number line *(Inv. 2)*	A line for representing and graphing numbers. Each point on the line corresponds to a number.  **number line**
recta numérica	Recta para representar y graficar números. Cada punto de la recta corresponde a un número.

number sentence *(maintained)*	A complete sentence that uses numbers and symbols instead of words. *See also* **equation.** *The **number sentence** 4 + 5 = 9 means "four plus five equals nine."*

enunciado numérico

Un enunciado completo que usa números y símbolos en lugar de palabras. *Ver también* **ecuación.**

*El **enunciado numérico** 4 + 5 = 9 significa "cuatro más cinco es igual a nueve".*

---

numerator
(19)

The top number of a fraction; The number that tells how many parts of a whole are counted.

*The **numerator** of the fraction is 1. One part of the whole circle is shaded.*

numerador

El término superior de una fracción. El número que nos dice cuantas partes de un entero se cuentan.

*El **numerador** de la fracción es 1.*

Una parte del círculo completo esta sombreada.

O

---

obtuse angle
(17)

An angle whose measure is more than 90° and less than 180°.

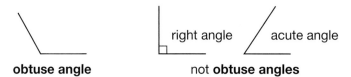

*An **obtuse angle** is larger than both a right angle and an acute angle.*

ángulo obtuso

Ángulo que mide más de 90° y menos de 180°.

*Un **ángulo obtuso** es más grande que un ángulo recto y que un ángulo agudo.*

---

obtuse triangle
(81)

A triangle whose largest angle measures more than 90° and less than 180°.

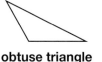
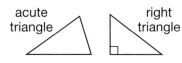

triángulo obtusángulo

Triángulo cuyo ángulo mayor mide más que 90° y menos que 180°.

---

octagon
(20)

A polygon with eight sides.

octágono

Un polígono con ocho lados.

---

odd numbers
(1)

Numbers that have a remainder of 1 when divided by 2; The numbers in this sequence: 1, 3, 5, 7, 9, 11, ...

***Odd numbers** have 1, 3, 5, 7, or 9 in the ones place.*

---

números impares

Números que cuando se dividen entre 2 tienen residuo de 1; los números en esta secuencia: 1, 3, 5, 7, 9, 11....

*Los **números impares** tienen 1, 3, 5, 7, ó 9 en el lugar de las unidades.*

---

order of operations
(9)

The set of rules for the order in which to solve math problems.

*Following the **order of operations**, we multiply and divide within an expression before we add and subtract.*

orden de operaciones

El conjunto de reglas del orden para resolver problemas matemáticos.

*Siguiendo el **orden de operaciones** multiplicamos y dividimos dentro de la expresión antes de sumar y restar.*

---

ordinal numbers
(maintained)

Numbers that describe position or order.

*"First," "second," and "third" are **ordinal numbers.***

números ordinales

Números que describen posición u orden.

*"Primero", "segundo" y "tercero" son **números ordinales.***

---

origin
(Inv. 7)

(1) The location of the number 0 on a number line.

origin on a number line

(2) The point (0, 0) on a coordinate plane.

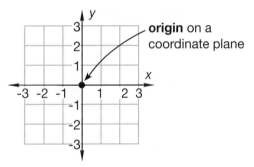

origin on a coordinate plane

origen

(1) La ubicación del número 0 en una recta numérica.

(2) El punto (0, 0) en un plano coordenado.

---

ounce
(80)

A unit of weight in the customary system. It is also a measure of capacity. *See also **fluid ounce.***

*Sixteen **ounces** equals a pound. Sixteen **fluid ounces** equals a pint.*

onza

Una unidad de peso en el sistema usual. También es una medida de capacidad. *Ver también **onza líquida.***

*Dieciseis **onzas** es igual a una libra. Dieciseis **onzas líquidas** es igual a una pinta.*

---

outcome
(110)

The end result of a probability experiment.

resultado

El resultado final de un experimento de probabilidad.

outlier *(Inv. 6)*	A number in a list of data that is distant from the other numbers. *1, 5, 4, 3, 6, 28, 7, 2* *In the data at right, the number 28 is an **outlier** because it is distant from the other numbers in the list.*
valor lejano	Un número en una lista de datos que es distante de los demás números en la lista. *En los datos a la derecha el número 28 es un **valor extremo**, porque su valor es mayor que el de los demás números de la lista.*

P

parallel lines *(17)*	Lines that stay the same distance apart; lines that do not cross  **parallel lines**
rectas paralelas	Rectas que permanecen separadas a la misma distancia y que nunca se cruzan.

parallelogram *(90)*	A quadrilateral that has two pairs of parallel sides.  **parallelograms** not a **parallelogram**
paralelogramo	Cuadrilátero que tiene dos pares de lados paralelos.

parentheses *(9)*	A pair of symbols used to separate parts of an expression so that those parts may be evaluated first: ( ) $15 - (12 - 4)$ *In the expression $15 - (12 - 4)$, the **parentheses** indicate that $12 - 4$ should be calculated before subtracting the result from 15.*
paréntesis	Un par de símbolos que se usan para separar partes de una expresión para que esas partes puedan ser evaluadas primero. $15 - (12 - 4)$ *En la expresión $15 - (12 - 4)$ el **paréntesis** indica que $12 - 4$ debe ser calculado antes de restar el resultado de 15.*

partial product *(86)*	A product formed by multiplying one factor by one digit of a second factor when the second fact has more than one digit. $$\begin{array}{r} 53 \\ \times\ 26 \\ \hline 318 \\ +\ 106 \\ \hline 1378 \end{array}$$ partial products
producto parcial	Un producto que se forma multiplicando un factor por un dígito de un segundo factor cuando el segundo factor tiene más de un dígito.

pentagon *(20)*	A polygon with five sides.

pentagon

pentágono	Un polígono con cinco lados.

per *(58)*	A term that means "in each." *A car traveling 50 miles **per** hour (50 mph) is traveling 50 miles in each hour.*
por cada	Un término que significa "en cada". *Un carro viajando 50 millas por hora (50 mph) está viajando 50 millas **por cada** hora.*

perfect square *(Inv. 3)*	*See **square number**.*
cuadrado perfecto	*Ver **número al cuadrado**.*

perimeter *(20)*	The distance around a closed, flat shape.

*The **perimeter** of this rectangle (from point A around to point A) is 32 inches.*

perímetro	Distancia alrededor de una figura cerrada y plana. *El **perímetro** de este rectángulo (desde el punto A alrededor del rectángulo hasta el punto A) es 32 pulgadas.*

perpendicular lines *(17)*	Two lines that intersect at right angles.

perpendicular lines   not **perpendicular lines**

rectas perpendiculares	Dos rectas que intersecan en ángulos rectos.

pictograph
(Inv. 5)

A graph that uses symbols to represent data.

Stars We Saw	
Tom	☆ ☆ ☆ ☆
Bob	☆ ☆
Sue	☆ ☆ ☆ ☆
Ming	☆ ☆ ☆ ☆ ☆
Juan	☆ ☆ ☆ ☆ ☆ ☆

*This is a **pictograph**. It shows how many stars each person saw.*

pictograma

Gráfica que utiliza símbolos para representar datos.

*Éste es un **pictograma**. Muestra el número de estrellas que vio cada persona.*

---

pie graph
(Inv. 5)

*See **circle graph**.*

diagrama circular

*Ver **gráfica circular**.*

---

place value
(2)

The value of a digit based on its position within a number.

341
23
+ 7
___
371

Place value tells us that 4 in 341 is worth "4 tens." In addition problems we align digits with the same **place value**.

valor posicional

Valor de un dígito de acuerdo al lugar que ocupa en el número.

341
23
+ 7
___
371

El **valor posicional** indica que el 4 en 341 vale "cuatro decenas". En los problemas de suma y resta, se alinean los dígitos que tienen el mismo **valor posicional**.

---

plane
(109)

A two-dimensional, flat surface that continues in all directions without end.

plano

Una superficie plana de dos dimensiones que continúa en todas direcciones sin fin.

---

p.m.
(13)

The period of time from noon to just before midnight.

*I go to bed at 9 **p.m.**, which is 9 o'clock at night.*

p.m.

Período de tiempo desde el mediodía hasta justo antes de la medianoche.

*Me voy a dormir a las 9 **p.m.**, lo cual es las 9 de la noche.*

---

point
(17)

An exact position.

•A   *This dot represents **point** A.*

punto

Una posición exacta.

*Esta marca representa el **punto** A.*

polygon
(20)

A closed, flat shape with straight sides.

polygons   not **polygons**

polígono

Figura cerrada y plana que tiene lados rectos.

---

population
(Inv. 6)

A group of people about whom information is gathered during a survey.

*A soft drink company wanted to know the favorite beverage of people in Indiana. The **population** they gathered information about was the people of Indiana.*

población

Un grupo de gente de la cual se obtiene información durante una encuesta.

*Una compañía de sodas quería saber cuál es la bebida favorita de la gente en Indiana. La **población** de la cual recolectaron información fue la gente de Indiana.*

---

positive numbers
(Inv. 2)

Numbers greater than zero.

*0.25 and 157 are **positive numbers.***

*−40 and 0 are not **positive numbers.***

números positivos

Números mayores que cero.

*0.25 y 157 son **números positivos.***

*−40 y 0 no son **números positivos.***

---

pound
(80)

A customary measurement of weight.

*One **pound** is 16 ounces.*

libra

Una medida usual de peso.

*Una **libra** es igual a 16 onzas.*

---

prime number
(55)

A counting number greater than 1 whose only two factors are the number 1 and itself.

*7 is a **prime number.** Its only factors are 1 and 7.*

*10 is not a **prime number.** Its factors are 1, 2, 5, and 10.*

número primo

Número de contes mayor que 1, cuyos dos únicos factores son el 1 y el propio número.

*7 es un **número primo.** Sus únicos factores son 1 y 7.*

*10 no es un **número primo.** Sus factores son 1, 2, 5 y 10.*

---

prism
(108)

*See **geometric solid.***

prisma

*Ver **sólido geométrico.***

---

probability
(110)

A way of describing the likelihood of an event; the ratio of favorable outcomes to all possible outcomes

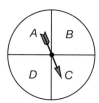

*The **probability** of the spinner landing on C is $\frac{1}{4}$.*

probabilidad	Manera de describir la ocurrencia de un suceso; la razón de resultados favorables a todos los resultados posibles. *La **probabilidad** de obtener 3 al lanzar un cubo estándar de números es $\frac{1}{6}$.*

---

product (23)	The result of multiplication. $5 \times 3 = 15$ The **product** of 5 and 3 is 15.
producto	Resultado de una multiplicación. $5 \times 3 = 15$ El **producto** de 5 por 3 es 15.

---

proper fraction (89)	A fraction whose denominator is greater than its numerator. $\frac{3}{4}$ is a **proper fraction.** $\frac{4}{3}$ is not a **proper fraction.**
fracción propia	Una fracción cuyo denominador es mayor que el numerador. $\frac{3}{4}$ es una **fracción propia.** $\frac{4}{3}$ no es una **fracción propia.**

Property of Zero for Multiplication (23)	Zero times any number is zero. In symbolic form, $0 \times a = 0$. **The Property of Zero for Multiplication** *tells us that* $89 \times 0 = 0$.
propiedad del cero en la multiplicación	Cero multiplicado por cualquier número es cero. En forma simbólica, $0 \times a = 0$. *La **propiedad del cero en la multiplicación** dice que $89 \times 0 = 0$.*

---

pyramid (96, 109)	A three-dimensional solid with a polygon as its base and triangular faces that meet at the apex. pyramid
pirámide	Figura geométrica de tres dimensiones, con un polígono en su base y caras triangulares que se encuentran en un vértice.

Q

---

quadrilateral (20)	Any four-sided polygon. *Each of these polygons has 4 sides. They are all **quadrilaterals**.*
cuadrilátero	Cualquier polígono de cuatro lados. *Cada uno de estos polígonos tiene 4 lados. Todos son **cuadriláteros**.*

quarter *(19)*	A term that means one-fourth.
cuarto	Un término que significa un cuarto.
quarter turn *(78)*	A turn measuring 90°.
cuarto de giro	Un giro que mide 90°.
quotient *(54)*	The result of division.

$$12 \div 3 = 4 \qquad 3\overline{)12}^{\,4} \qquad \frac{12}{3} = 4$$

*The **quotient** is 4 in each of these problems.*

cociente	Resultado de una división.

*El **cociente** es 4 en cada una de estas operaciones.*

R

radius *(18)*	(Plural: *radii*) The distance from the center of a circle to a point on the circle.

1 cm

*The **radius** of this circle is 1 centimeter.*

radio	Distancia desde el centro de un círculo hasta un punto del círculo.

*El **radio** de este círculo mide 1 centímetro.*

range *(95)*	The difference between the largest number and smallest number in a list.

5, 17, 12, 34, 28, 13

*To calculate the **range** of this list, we subtract the smallest number from the largest number. The **range** of this list is 29.*

intervalo	Diferencia entre el número mayor y el número menor de una lista.

5, 17, 12, 34, 28, 13

*Para calcular el **intervalo** de esta lista, se resta el número menor del número mayor. El **intervalo** de esta lista es 29.*

rate *(58)*	A measure of how far or how many are in one time group.

*The leaky faucet wasted water at the **rate** of 1 liter per day.*

tasa	Una medida de cuánto hay en un grupo por unidad de tiempo.

*La llave de agua con fuga desperdiciaba agua a una **tasa** de 1 litro al día.*

ray *(17)*	A part of a line that begins at a point and continues without end in one direction.

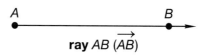

ray AB ($\overrightarrow{AB}$)

rayo   Parte de una recta que empieza en un punto y continúa indefinidamente en una dirección.

---

rectangle
(18)

A quadrilateral that has four right angles.

rectangles   not **rectangles**

rectángulo   Cuadrilátero que tiene cuatro ángulos rectos.

---

rectangular prism
(96)

A geometric solid with 6 rectangular faces.

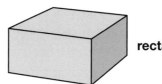

rectangular prism

prisma rectangular   Un sóldio geométrico con 6 caras rectangulares.

---

reduce
(Inv. 8)

To rewrite a fraction in lowest terms.

*If we **reduce** the fraction $\frac{9}{12}$, we get $\frac{3}{4}$.*

reducir   Escribir una fracción a su mínima expresión.

*Si **reducimos** $\frac{9}{12}$, obtenemos $\frac{3}{4}$.*

---

reflective symmetry
(82)

A figure has *reflective symmetry* if it can be divided into two halves that are mirror images of each other. *See also* **line of symmetry.**

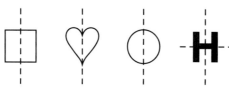

These figures have **reflective symmetry.**   These figures do not have **reflective symmetry.**

simetría de reflexión   Una figura tiene *simetría de reflexión* si puede ser dividida en dos mitades una de las cuales es la imagen espejo de la otra. *Ver también* **eje de simetría.**

---

regrouping
(maintained)

To rearrange quantities in place values of numbers during calculations.

$$214 \longrightarrow \overset{1\ 10\ 14}{\cancel{2}\cancel{1}\cancel{4}}$$
$$-\ 39 \qquad\qquad -\ 39$$
$$\overline{\qquad\qquad\quad 175}$$

*Subtraction of 39 from 214 requires **regrouping.***

reagrupar   Reordenar cantidades según los valores poscionales de números al hacer cálculos.

*La resta de 39 de 214 requiere **reagrupación.***

---

regular polygon
(20)

A polygon in which all sides have equal lengths and all angles have equal measures.

regular polygons   **not regular polygons**

polígono regular

Polígono en el cual todos los lados tienen la misma longitud y todos los ángulos tienen la misma medida.

---

remainder
(54)

An amount that is left after division.

$$7 \ R \ 1$$
$$2\overline{)15}$$
$$\underline{14}$$
$$1$$

*When 15 is divided by 2, there is a **remainder** of 1.*

residuo

Cantidad que queda después de dividir.

$$7 \ R \ 1$$
$$2\overline{)15}$$
$$\underline{14}$$
$$1$$

*Cuando se divide 15 entre 2, queda **residuo** 1.*

---

rhombus
(90)

A parallelogram with all four sides of equal length.

rhombuses   **not rhombuses**

rombo

Paralelogramo con sus cuatro lados de igual longitud.

---

right angle
(17)

An angle that forms a square corner and measures 90°. It is often marked with a small square.

obtuse angle   acute angle

right angle   **not right angles**

*A **right angle** is larger than an acute angle and smaller than an obtuse angle.*

ángulo recto

Ángulo que forma una esquina cuadrada y mide 90°. Se indica con frecuencia con un pequeño cuadrado.

*Un **ángulo recto** es mayor que un ángulo agudo y más pequeño que un ángulo obtuso.*

right triangle *(81)*	A triangle whose largest angle measures 90°. right triangle acute triangle obtuse triangle **right triangle** not **right triangles**
triángulo rectángulo	Triángulo cuyo ángulo mayor mide 90°.
rotational symmetry *(82)*	A figure has *rotational symmetry* if it can be rotated less than a full turn and appear in its original orientation. **S △ ❁ Z** ▱ **M** ⌂ These figures have **rotational symmetry.** These figures do not have **rotational symmetry.**
simetría de rotación	Una figura tiene *simetría de rotación* si puede ser rotada menos que un giro completo y aparecer en su orientación original.
round *(22)*	To express a calculation or measure to a specific degree of accuracy. *To the nearest hundred dollars, $294 **rounds** to $300.*
redondear	Expresar un cálculo o medida hasta cierto grado de precisión. *A la centena de dólares más cerca, $294 se **redondea** a $300.*

S

sales tax *(87)*	The tax charged on the sale of an item and based upon the item's purchase price. *If the **sales-tax** rate is 8%, the **sales tax** on a $5.00 item will be $5.00 × 8% = $0.40.*
impuesto sobre la venta	Impuesto que se carga al vender un objeto y que se calcula como un porcentaje del precio del objeto. *Si la tasa de impuesto es 8%, el **impuesto sobre la venta** de un objeto que cuesta $5.00 es: $5.00 × 8% = $0.40.*
sample *(Inv. 6)*	A part of a population used to conduct a survey. *Mya wanted to know the favorite television show of the fourth-grade students at her school. She asked only the students in Room 3 her survey question. In her survey, the population was the fourth-grade students at the school, and the **sample** was the students in Room 3.*
muestra	Una parte de una población que se usa para realizar una encuesta. *Mya quería saber cuál es el programa favorito de los estudiantes de cuarto grado de su escuela. Ella hizo la pregunta de su encuesta a sólo el Salón 3. En su encuesta, la población era los estudiantes del cuarto grado de su escuela, y su **muestra** fue los estudiantes del Salón 3.*

| **scale** | A type of number line used for measuring. |
| (21) | |

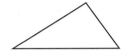

*The distance between each mark on this ruler's **scale** is 1 centimeter.*

escala Un tipo de recta númerica que se usa para medir.

*La distancia entre cada marca en la **escala** de esta regla es 1 centímetro.*

| **scalene triangle** | A triangle with three sides of different lengths. |
| (81) | |

*All three sides of this **scalene triangle** have different lengths.*

triángulo escaleno Triángulo con todos sus lados de diferente longitud.

*Los tres lados de este **triángulo escaleno** tienen diferente longitud.*

| **schedule** | A list of events organized by the times at which they are |
| (maintained) | planned to occur. |

Sarah's Class Schedule

8:15 a.m.	Homeroom
9:00 a.m.	Science
10:15 a.m.	Reading
11:30 a.m.	Lunch and recess
12:15 p.m.	Math
1:30 p.m.	English
2:45 p.m.	Art and music
3:30 p.m.	End of school

calendario, horario Una lista de sucesos organizados según la hora cuando están planeados.

| **sector** | A region bordered by part of a circle and two radii. |
| (Inv. 11) | |

*This circle is divided into 3 **sectors.** One **sector** of the circle is shaded.*

sector Región de un círculo limitada por un arco y dos radios.

*Este círculo esta dividido en 3 **sectores.** Un **sector** del círculo está sombreado.*

| **segment** | *See **line segment.*** |
| (17) | |

segmento *Ver **segmento de recta.***

sequence (1)	A list of numbers arranged according to a certain rule. *The numbers 5, 10, 15, 20, ... form a **sequence.** The rule is "count up by fives."*
secuencia	Lista de números ordenados de acuerdo a una regla. *Los números 5, 10, 15, 20, ... forman una **secuencia.** La regla es "contar hacia adelante de cinco en cinco".*

side (maintained)	A line segment that is part of a polygon. *The arrow is pointing to one side. This pentagon has 5 **sides.***
lado	Segmento de recta que forma parte de un polígono. *Este pentágono tiene 5 **lados.***

similar (70)	Having the same shape but not necessarily the same size. Matching angles of *similar* figures are equal. 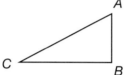 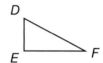 *△ABC and △DEF are **similar.** They have the same shape, but not the same size.*
semejante	Que tiene la misma forma, pero no necesariamente el mismo tamaño. Las dimensiones de figuras semejantes son proporcionales. *△ABC y △DEF son **semejantes.** Tienen la misma forma, pero diferente tamaño.*

solid (96)	*See **geometric solid.***
sólido	*Ver **sólido geométrico.***

sphere (96)	A round geometric solid having every point on its surface at an equal distance from its center. sphere
esfera	Un sólido geométrico redondo que tiene cada punto de su superficie a la misma distancia de su centro.

square
(18)

1. A rectangle with all four sides of equal length.

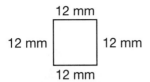

*All four sides of this **square** are 12 millimeters long.*

2. The product of a number and itself.

*The **square** of 4 is 16.*

cuadrado

1. Un rectángulo con sus cuatro lados de igual longitud.
   *Los cuatro lados de este **cuadrado** miden 12 milímetros.*

2. El producto de un número por sí mismo.
   *El **cuadrado** de 4 es 16.*

---

square centimeter
(Inv. 3)

A measure of area equal to that of a square with 1-centimeter sides.

centímetro cuadrado   Medida de un área igual a la de un cuadrado con lados de 1 centímetro.

---

square foot
(Inv. 3)

A unit of area equal to a square with 1-foot sides.

pie cuadrado   Unidad de área igual a un cuadrado con lados que miden un pie de longitud.

---

square inch
(Inv. 3)

A measure of area equal to that of a square with 1-inch sides.

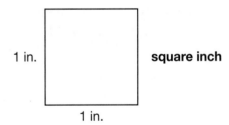

pulgada cuadrada   Medida de un área igual a la de un cuadrado con lados de 1 pulgada.

---

square number
(25)

The product when a whole number is multiplied by itself.

*The number 9 is a **square number** because $9 = 3^2$.*

número al cuadrado

El producto de un número entero multiplicado por sí mismo.
*El número 9 es un **número al cuadrado** porque $9 = 3^2$.*

---

square root
(Inv. 3)

One of two equal factors of a number. The symbol for the principal, or positive, *square root* of a number is $\sqrt{}$.

*A **square root** of 49 is 7 because $7 \times 7 = 49$.*

raíz cuadrada

Uno de dos factores iguales de un número. El símbolo de la *raíz cuadrada* de un número es $\sqrt{}$, y se le llama *radical*.
*La **raíz cuadrada** de 49 es 7, porque $7 \times 7 = 49$.*

square unit *(Inv. 3)*	An area equal to the area of a square with sides of designated length.

> *The shaded part is 1 **square unit**. The area of the large rectangle is 8 **square units**.*

unidad cuadrada

Un área igual al área de un cuadrado con lados de una longitud designada.

> *La parte sombreda es 1 **unidad cuadrada**. El área del rectángulo grande es de 8 **unidades cuadradas**.*

---

subtraction
(maintained)

The arithmetic operation that reduces a number by an amount determined by another number

$15 - 12 = 3$ *We use **subtraction** to take 12 away from 15.*

resta

La operación aritmética que reduce un número por cierta cantidad determinada por otro número.

$15 - 12 = 3$ *Utilizamos la **resta** para quitar 12 de 15.*

---

sum
(maintained)

The result of addition.

$2 + 3 = 5$ *The **sum** of 2 and 3 is 5.*

suma

Resultado de una suma.

$2 + 3 = 5$ *La **suma** de 2 más 3 es 5.*

---

survey
(Inv. 6)

A method of collecting data about a particular population.

*Mia conducted a **survey** by asking each of her classmates the name of his or her favorite television show.*

encuesta

Método de reunir información acerca de una población en particular.

*Mia hizo una **encuesta** entre sus compañeros para averiguar cuál era su programa favorito de televisión.*

---

symmetry
(82)

Correspondence in size and shape on either side of a dividing line. This type of *symmetry* is known as *reflective symmetry. See also* **line of symmetry.**

These figures have **reflective symmetry.**   These figures do not have **reflective symmetry.**

simetría

Correspondencia en tamaño y forma a cada lado de una línea divisoria. Este tipo de *simetría* es conocida como *simetría de reflexión. Ver también* **eje de simetría.**

T

| table | A way of organizing data in columns and rows. |

Our Group Scores

Name	Grade
Group 1	98
Group 2	72
Group 3	85
Group 4	96

*This **table** shows the scores of four groups.*

tabla — Una manera organizada de datos en columnas y filas.
*Esta **tabla** muestra las calificaciones de cuatro grupos.*

---

tally mark
(Inv. 6)

A small mark used to help keep track of a count.

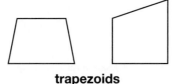

*I used **tally marks** to count cars.*
I counted five cars.

marca de conteo — Una pequeña marca que se usa para llevar la cuenta.
*Usé **marcas de conteo** para contar carros. Yo conté cinco carros.*

---

tenth
(Inv. 4)

One out of ten parts, or $\frac{1}{10}$.
*The decimal form of one **tenth** is 0.1.*

décimo(a) — Una de diez partes ó $\frac{1}{10}$.
*La forma decimal de un **décimo** es 0.1.*

---

tick mark
(Inv. 2)

A mark dividing a number line into smaller portions.

marca de un punto — Una marca que divide a una recta numérica en partes más pequeñas.

---

ton
(80)

A customary measurement of weight equal to 2000 pounds.

tonelada — Una medida usual de peso.

---

trapezoid
(90)

A quadrilateral with exactly one pair of parallel sides.

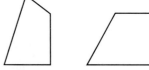

trapezoids   not **trapezoids**

trapecio — Cuadrilátero que tiene exactamente un par de lados paralelos.

tree diagram *(72)*	A way to use branches to organize the choices of a combination problem.

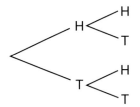

tree diagram

diagrama de árbol	Una manera de usar ramas para organizar los opciones de un problema de comparación.

triangle *(20)*	A polygon with three sides and three angles.

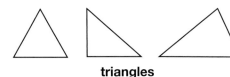

triangles

triángulo	Un polígono con tres lados y tres ángulos.

triangular prism *(96)*	A geometric solid with 2 triangular bases and 3 rectangular faces.

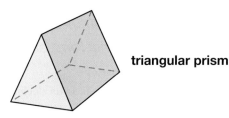

triangular prism

prisma triangular	Un sólido geométrico con 3 caras rectangulares y 2 bases triangulares.

U

unit *(maintained)*	Any standard object or quantity used for measurement. *Grams, pounds, liters, gallons, inches, and meters are all **units**.*
unidad	Cualquier objeto estándar o cantidad que se usa para medir. *Gramos, libras, galones, pulgadas y metros son **unidades**.*

U.S. Customary System *(42)*	A system of measurement used almost exclusively in the United States. *Pounds, quarts, and feet are units in the **U.S. Customary System**.*
Sistema usual de EE.UU.	Unidades de medida que se usan casi exclusivamente en EE.UU. *Libras, cuartos y pies son unidades del **Sistema usual de EE.UU.***

V

vertex
(17, 96)

(Plural: *vertices*) A point of an angle, polygon, or solid where two or more lines, rays, or segments meet.

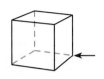

*The arrow is pointing to one **vertex** of this cube. A cube has eight **vertices.***

vértice

Punto de un ángulo, polígono o sólido, donde se unen dos o más rectas, semirrectas o segmentos de recta.

*La flecha apunta hacia un **vértice** de este cubo. Un cubo tiene ocho **vértices.***

vertical
(Inv. 7)

Upright; perpendicular to horizontal

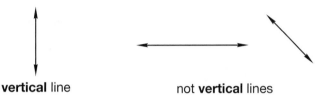

vertical line   not **vertical** lines

vertical

Hacia arriba; perpendicular a la horizontal.

W

weight
(80)

The measure of the force of gravity on an object. Units of *weight* in the customary system include ounces, pounds, and tons.

*The **weight** of a bowling ball would be less on the moon than on Earth because the force of gravity is weaker on the moon.*

peso

La medida de la fuerza de gravedad sobre un objeto. Las unidades de *peso* en el sistema usual incluyen onzas, libras y toneladas.

*El **peso** de una bola de boliche es menor en la Luna que en la Tierra porque la fuerza de gravedad es menor en la Luna.*

whole numbers
(3)

All the numbers in this sequence: 0, 1, 2, 3, 4, 5, 6, 7, 8, 9, ...

*The number 35 is a **whole number,** but $35\frac{1}{2}$ and 3.2 are not.* ***Whole numbers** are the counting numbers and zero.*

números enteros

Todos los números en esta secuencia: 0, 1, 2, 3, 4, 5, 6, 7, 8, 9 ....

*El número 35 es un **número entero** pero $35\frac{1}{2}$ y 3.2 no lo son.*
*Los **números enteros** son los números de conteo y el cero.*

width
(20)

The shorter dimension of a rectangle.

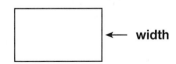

ancho

La dimensión más corta de un rectángulo.

X

x-axis
(Inv. 7)

The horizontal number line of a coordinate plane.

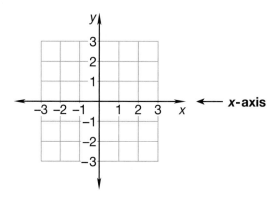

eje-x La recta numérica horizontal de un plano coordenado.

Y

y-axis
(Inv. 7)

The vertical number line of a coordinate plane.

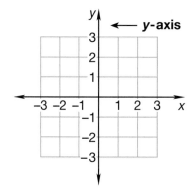

eje-y La recta numérica vertical de un plano coordenado.

yard
(42)

A customary measurement of length.

yarda Una medida usual de longitud.

Symbols

Symbol	Meaning	Example
$\triangle$	Triangle	$\triangle ABC$
$\angle$	Angle	$\angle ABC$
$\rightarrow$	Ray	$\overrightarrow{AB}$
$\leftrightarrow$	Line	$\overleftrightarrow{AB}$
$-$	Line segment	$\overline{AB}$
$\perp$	Perpendicular to	$AB \perp BC$
$\parallel$	Parallel to	$AB \parallel BC$
$<$	Less than	$2 < 3$
$>$	Greater than	$3 > 2$
$=$	Equal to	$2 = 2$
°F	Degrees Fahrenheit	100°F
°C	Degrees Celsius	32°C
$\llcorner$	Right angle (90° angle)	
…	And so on	1, 2, 3, …
$\times$	Multiply	9×3
$\cdot$	Multiply	$3 \cdot 3 = 9$
$\div$	Divide	$9 \div 3$
$+$	Add	$9 + 3$
$-$	Subtract	$9 - 3$
$\overline{)}$	Divided into	$3\overline{)9}$
R or r	Remainder	3 R 2
x^2	"x" squared (times itself)	$3^2 = 3 \times 3 = 9$
x^3	"x" cubed	$3^3 = 3 \times 3 \times 3 = 27$
$\sqrt{}$	Square root	$\sqrt{9} = 3$ because $3 \times 3 = 9$.

Abbreviations

Abbreviation	Meaning
ft	Foot
in.	Inch
yd	Yard
mi	Mile
m	Meter
cm	Centimeter
mm	Millimeter
km	Kilometer
L	Liter
ml or mL	Milliliter
lb	Pound
oz	Ounce
kg	Kilogram
g	Gram
mg	Milligram
qt	Quart
pt	Pint
c	Cup
gal	Gallon

Formulas

Purpose	Formula
Perimeter of a rectangle	$P = 2l + 2w$
Perimeter of a square	$P = 4s$
Area of a square	$A = s^2$
Area of a rectangle	$A = l \cdot w$
Volume of a cube	$V = s^3$
Volume of a rectangular prism	$V = l \cdot w \cdot h$

Símbolos/Signos

Símbolo/Signo	Significa	Ejemplo
$\triangle$	Triángulo	$\triangle ABC$
$\angle$	Ángulo	$\angle ABC$
$\rightarrow$	Rayo	$\overrightarrow{AB}$
$\leftrightarrow$	Recta	$\overleftrightarrow{AB}$
—	Segmento de recta	$\overline{AB}$
$\perp$	Perpendicular a	$AB \perp BC$
$\parallel$	Paralelo a	$AB \parallel BC$
$<$	Menor que	$2 < 3$
$>$	Mayor que	$3 > 2$
$=$	Igual a	$2 = 2$
°F	Grados Fahrenheit	100°F
°C	Grados Celsius	32°C
$\llcorner$	Ángulo recto (ángulo de 90°)	
…	Y más, etcétera	1, 2, 3, …
$\times$	Multiplica	9×3
$\cdot$	Multiplica	$3 \cdot 3 = 9$
$\div$	Divide	$9 \div 3$
$+$	Suma	$9 + 3$
$-$	Resta	$9 - 3$
$\overline{)}$	Dividido entre	$3\overline{)9}$
R or r	Residuo	3 R 2
x^2	"x" al cuadrado (por sí mismo)	$3^2 = 3 \times 3 = 9$
x^3	"x" al cubo	$3^3 = 3 \times 3 \times 3 = 27$
$\sqrt{}$	Raíz cuadrada	$\sqrt{9} = 3$ por que $3 \times 3 = 9$.

Abreviaturas

Abreviatura	Significa
pie	pie
pulg	pulgada
yd	yarda
mi	milla
m	metro
cm	centímetro
mm	milímetro
km	kilómetro
L	litro
mL	mililitro
lb	libra
oz	onza
kg	kilogramo
g	gramo
mg	miligramo
ct	cuarto
pt	pinta
tz	taza
gal	galón

Fórmulas

Propósito	Fórmula
Perímetro de un rectángulo	$P = 2L + 2a$
Perímetro de un cuadrado	$P = 4l$
Área de un cuadrado	$A = l^2$
Área de un rectángulo	$A = L \cdot a$
Volumen de un cubo	$V = l^3$
Volumen de un prisma rectangular	$V = L \cdot a \cdot h$

A

Abbreviations. *See also* **Symbols and signs**
  a.m., 74
  Celsius (°C), 136
  centimeter (cm), 282–283
  cup (c), 488
  Fahrenheit (°F), 136
  fluid ounce (fl oz), 488
  foot (ft), 283
  gallon (gal), 488
  grams (g), 533
  inch (in.), 283
  kilograms (kg), 533
  kilometer (km), 283
  liter (L), 489
  meter (m), 283
  mile (mi), 283
  millimeter (mm), 282–283
  ounce (oz), 532
  pint (pt), 488
  p.m., 74
  pound (lb), 532
  quart (qt), 488
  yard (yd), 283

Act it out, 5

Activities
  arrays to find factors, 371–372
  collecting data, 629
  comparing fractions, 382–383
  congruence and rotations, 521
  constructing prisms, 640–641
  degrees and rotations, 518–519
  different areas, 206
  different perimeters, 124–125
  displaying information on graphs, 338–340
  drawing a circle, 111
  drawing number lines, 130
  estimating perimeter and area, 202
  finding perimeter and area, 202
  finding time, 76
  fraction manipulatives, 537–538
  geometric solids in the real world, 635–636
  graphing pay rates, 666
  improper fractions and mixed numbers, 588
  measuring temperature, 137
  mixed numbers and improper fractions, 588
  models of pyramids, 716–717
  multiplication table to divide, 227
  perimeter and area, 202
  probability experiments, 728–730
  quadrilaterals and symmetry, 596
  quadrilaterals in the classroom, 596
  real-world segments and angles, 104

  reflections and lines of symmetry, 549–550
  relating fractions and decimals, 538–539
  rotations and congruence, 521
  rotations and degrees, 518–519
  similarity and congruence, 464
  symmetry and quadrilaterals, 596
  writing word problems, 61

Acute angles, 103–105

Acute triangles, 542–543

Addends, missing, 23–24

Addition
  algorithm, 36–37
  Associative Property of, 36, 168
  checking answers by, 84
  columns of numbers, 63–64
  Commutative Property of, 36
  of decimals, 299–301
  of fractions, 738
  of fractions with common denominators, 660–661
  inverse operations, 42–43, 54, 84, 183
  large numbers, 344–345
  mixed numbers, 661
  of money amounts, 32–33, 344–345
  with regrouping, 32–33, 63–65
  repeated, 258
  "some plus some more" formula, 68–70
  of three-digit numbers, 31–33
  word problems, 69

Algorithms
  addition, 36–37
  division, 485–486
  multiplication, 150–153
  subtraction, 58

a.m., 74

Angles
  acute, 103–105
  defined, 103
  obtuse, 103–105
  in parallelograms, 594
  in quadrilaterals, 595
  right, 103–105

Approximation. *See* **Estimation**

Area
  activities, 202
  of complex figures, 748–749
  models, 198–204, 371–372
  and perimeter, 206–208
  of rectangles, 442–444, 649

Arithmetic answers, estimation, 410–412

Arithmetic operations. *See* **Addition; Division; Multiplication; Subtraction**

Arrays, 198, 371–372

Associative Property
   of Addition, 168
   of Multiplication, 168–169

B

Bar graphs
   defined, 335
   display data using, 336
   making, 507–508, 339

Base, 437. *See also* **Exponents**

Base-ten system, 271

Bases of geometric solids, 710, 712

C

Calculation (mental math), 233, 239, 247, 253, 259,
   265, 275, 281, 287, 293, 299, 305, 311, 318, 323,
   329, 343, 349, 355, 361, 369, 376, 381, 387, 393,
   399, 409, 417, 423, 431, 436, 441, 447, 451, 457,
   463, 473, 479, 487, 493, 499, 505, 513, 517, 525,
   531, 541, 547, 555, 561, 567, 572, 577, 581, 586,
   593, 603, 609, 615, 622, 627, 633, 639, 646, 653,
   659, 669, 674, 681, 689, 695, 700, 705, 710, 715,
   721, 732, 737, 742, 747

Calculators, 350, 607

Capacity, 487–489

Celsius (˚C), 136

Center, 111

Centimeter (cm), 282

Cents and dollars, 181–184. *See also* **Money**

Chance. *See* **Probability**

Checking answers. *See* **Inverse operations**

Circle graphs
   defined, 335
   display data using, 338
   making, 340

Circles
   activity, 111
   concepts of, 111
   as plane figures, 716

Classification
   of geometric solids, 634–636
   of polygons, 123
   of quadrilaterals, 594–596
   of triangles, 542–543

Clockwise, 518–520

Columns of numbers, 63–64

Combining, word problems about, 59, 67–70. *See also*
   Addition

Commas, 174, 324

Common denominators. *See also* **Denominators**
   addition and subtraction of fractions, 660–661
   renaming, 743–744

Common factors. *See* **Factors**

Communication
   discuss, 20, 23, 33, 37, 48, 54, 77, 143, 144, 150,
      242, 248, 289, 294, 320, 350, 351, 407, 408,
      425, 443, 465, 471, 472, 494, 507, 514, 519,
      534, 589, 601, 602, 604, 611, 617, 624, 628,
      629, 641, 655, 661, 733
   explain, 16, 22, 30, 34, 55, 57, 66, 72, 88, 106,
      114, 119, 120, 125, 126, 139, 146, 159, 160,
      179, 185, 186, 190, 191, 209, 221, 237, 243,
      245, 250, 251, 257, 261, 262, 263, 290, 297,
      303, 308, 315, 316, 328, 334, 335, 336, 337,
      338, 353, 358, 360, 365, 373, 378, 385, 395,
      402, 405, 408, 411, 418, 426, 427, 428, 429,
      433, 440, 445, 456, 462, 466, 467, 470, 483,
      484, 486, 491, 492, 496, 509, 515, 522, 535,
      544, 566, 569, 571, 583, 597, 598, 600, 607,
      613, 614, 620, 625, 637, 643, 656, 658, 663,
      668, 672, 676, 680, 691, 692, 697, 704, 713,
      726, 727, 728, 734, 741
   formulate, 10, 15, 22, 25, 28, 44, 50, 55, 59, 60,
      61, 65, 70, 71, 78, 84, 86, 93, 98, 99, 105, 108,
      112, 114, 117, 119, 125, 138, 140, 141, 145,
      148, 153, 158, 160, 164, 170, 172, 177, 184,
      190, 195, 209, 216, 228, 238, 243, 250, 257,
      261, 263, 264, 268, 270, 277, 280, 285, 290,
      296, 301, 314, 316, 320, 326, 347, 353, 357,
      360, 365, 373, 378, 389, 390, 396, 401, 404,
      405, 426, 433, 434, 435, 462, 492, 508, 522,
      530, 564, 569, 589, 637, 655, 666, 668, 672,
      676, 677, 686, 691

Commutative Property
   of Addition, 36
   of Multiplication, 152

Comparing
   decimals, 276
   equivalent decimals, 295
   estimates, 410
   fractions, 189, 648
   hundred thousands, 175
   number lines for, 132
   numbers in millions, 313, 325
   and ordering fractions, 382–384
   temperatures, 136–137
   by using place value, 20
   word problems about, 61, 96–98
   words to write comparisons, 133

Compasses, 111

Compatible numbers, 249, 501. *See also* **Estimation**

Complex figures, 748–749

Composite numbers, 377–378

Cones, 716

Congruence
   of geometric figures, 383, 463–465
   of prisms, 711–712
   rotations and, 521

Connections, 27, 28, 90, 124, 136, 157, 162, 168, 193, 214, 226, 248, 266, 306, 324, 338, 350, 369, 394, 411, 432, 458, 463, 474, 594, 601, 611, 617, 647, 666, 682, 716. *See also* **Math and Other Subjects; Math-to-Math Connections; Real-World Connections**

Coordinates
locating on a graph, 468–472
subtracting, 598–602

Coordinate plane, 468–472

Counterclockwise, 518–520

Counting numbers, 8

Cubed numbers, 437–438

Cubes, 716

Cup (c), 488

D

Data. *See also* **Graphs**
collecting with surveys, 404–408
displaying using graphs, 335–342
line plots and, 407
median, 407, 627–628
mode, 406, 628–629
range, 628

Decagons, 331

Decimal numbers. *See* **Decimals**

Decimal points
aligning, 299–301
"and" in reading or writing, 273–274
and cent sign, 181–184
in decimal division, 527
money and, 183
place values, 276–277

Decimals
comparing and ordering, 611–612
equivalent, 294–295, 418
place value, 276–277
relating to fractions, 271–274

Degrees, 518–520

Denominators. *See also* **Common denominators**
defined, 115
reducing fractions, 700–702

Diagrams. *See* **Draw a Picture or Diagram; Graphs**

Diameter, 111

Difference. *See also* **Subtraction**
finding for three-digit numbers, 54–55
larger-smaller, 83, 96–98

Digits
defined, 9
writing whole numbers, 18–19

Discuss. *See* **Communication**

Distance. *See* **Length**

Distributive Property, 674–675

Division
algorithm, 483–484
and multiplication answers, estimating, 604–605
answers ending with zeros, 500–501
fraction representation of, 120
multiplication as inverse of, 226, 234–235
with three-digit answers, 526–527, 556–557
with two-digit answers, 452–454, 457–460, 481
word problems, 356–357, 452–454, 457–459, 481–482, 501, 527
by zero, 232

Dollars. *See also* **Money**
and cents, 181–184
fractions of, 188–190
rounding to nearest, 306–307

Double bar graphs, 507–508

Draw a picture or diagram, 5

Drawing, to compare fractions, 383. *See also* **Graphs**

E

Early Finishers. *See* **Enrichment**

Edges, 635

Elapsed-time, 74–77

Elevens, facts, 194

Endpoints
as coordinates on a graph, 600
of segments, 102

Enrichment
Early Finishers, 52, 80, 108, 148, 212, 231, 246, 264, 270, 280, 304, 348, 386, 392, 416, 467, 478, 498, 512, 554, 608, 673, 679, 720, 746
Investigations, 59–62, 128–134, 198–204, 271–274, 335–342, 404–408, 468–472, 537–540, 600–602, 665–668, 727–731

Equal groups
division, 452
rate word problems formula, 388–389
word problems about, 60–61

Equations. *See also* **Representation**
graphing, 470–472
solving graphing relationships of, 665–668
two-step, 432–433, 616–618

Equilateral triangles, 110, 542–543

Equivalent decimals
concepts of, 294–295
rounding to nearest tenth, 418

Equivalent fractions
concepts of, 682–684
finding common denominators, 688
fraction manipulatives, 537–539

Estimation
area and perimeter, 202, 649
arithmetic answers, 410–412

Logical reasoning. *See* **Use Logical Reasoning**

M

Make an Organized List, 5

Make Generalizations, 121–122

Make it Simpler, 5

Make or Use a Table, Chart, or Graph, 5

Mass *vs.* **weight,** 532–534

Math and Other Subjects

Math and History, 140, 177, 190, 327, 333, 346, 358, 375, 569, 579, 619, 657

Math and Geography, 30, 56, 72, 86, 99, 100, 112, 119, 166, 172, 175, 197, 209, 211, 212, 224, 238, 250, 257, 264, 268, 298, 303, 312, 346, 348, 353, 366, 554, 563, 589, 624, 630, 658, 678, 707, 735, 740, 751

Math and Science, 44, 62, 65, 148, 177, 211, 228, 231, 261, 264, 292, 298, 309, 366563, 599, 623, 624, 664, 719

Math and Sports, 39, 65, 80, 84, 95, 138, 139, 164, 172, 190, 211, 218, 246, 283, 304, 309, 356, 360, 400, 543, 558, 564, 571, 58, 590, 591, 599, 605, 613, 619, 637, 638, 662, 684, 685, 686, 691, 692, 698, 707, 713, 741

Math Language, 14, 24, 42, 47, 48, 110, 111, 116, 142, 150, 168, 206, 226, 283, 313, 370, 388, 488, 506, 587, 640, 682, 695, 716, 744

Measurement. *See also* **Units of measure**

of area, 198–204

of length, 281–284

linear, 206

mental math, 141, 149, 156, 161, 173, 180, 187, 193, 205, 213, 219, 225, 233, 239, 247, 259, 265, 275, 281, 287, 293, 299, 305, 323, 329, 343, 349, 369, 381, 387, 393, 417, 431, 457, 463, 493, 499, 505, 513, 517, 531, 541, 577, 581, 586, 615, 622, 627, 695, 747

metric system, 283, 488

perimeter, 122–125, 202, 206–208, 284, 649, 748–749

of temperature, 135–138

of turns, 518–521

Measures of central tendency. *See* **Median; Mode; Range**

Median, 407, 627–628

Memory group, 193–194

Mental Math. *See* **Calculation; Estimation; Fractional Part; Geometry; Measurement; Money; Number Sense; Percent; Powers and roots; Probability; Review; Simplify; Time**

Meter (m), 283

Metric system, 283, 488, 533. *See also* **Units of measure**

Mile (mi), 283

Milliliter (mL), 488

Millimeter (mm), 282–283

Millions

reading and writing, 312–313

rounding, 319, 690

Mirror image, 548–549. *See also* **Symmetry**

Missing numbers. *See also* **Unknown numbers**

addends, 42–43

inverse operations, 42

in subtraction, 37–39, 42

Mixed numbers. *See also* **Improper fractions**

addition of, 661, 738

changing improper fractions to, 654

defined, 661

and fractions, 213–214

and improper fractions, 587–589

subtraction of, 661, 738

Mode, 406, 628–629

Models, area, 198–204. *See also* **Representation**

Money

addition of, 32–33, 344–345

in division, 527

estimation of amount, 411–412

finding information to solve problems, 506–507

formulas, 670

mental math, 67, 73, 81, 89, 95, 101, 109, 115, 121, 135, 141, 149, 156, 161, 167, 173, 180, 187, 193, 213, 233, 239, 253, 259, 265, 275, 281, 287, 293, 299, 305, 318, 323, 329, 349, 355, 361, 369, 376, 381, 387, 393, 399, 409, 417, 423, 431, 436, 441, 451, 463, 473, 479, 487, 499, 505, 513, 517, 541, 555, 567, 572, 586, 593, 603, 609, 627, 633, 639, 646, 653, 659, 669, 674, 681, 695, 700, 715, 721, 737

mixed numbers for, 182

multiplication of, 395, 475, 706

place value shown with, 13–14

rounding, 143, 307, 317, 345, 418–419

subtraction of, 83

word problems, 69–70

writing amounts of, 182

Multiples

of 10, 473–475

of 10 and 100, 248–249, 568–569

of 10, 11, and 12, 194

and factors, 369–372

Multiplication. *See also* **Exponents**

by 10, 100, and 1000, 562–563

algorithm, 258

Associative Property of, 168–169

checking answers, 459

Commutative Property of, 152

concept of, 150–152

defined, 150

Distributive Property of, 674–675

and division answers, estimating, 604–605

division as inverse of, 226–228, 234–235

expanded form, 258

facts, 151, 157, 162–164, 193–194

facts for 0's, 1's, 2's, and 5's, 157

facts for 9's and squares, 162–164

Identity Property of, 152
of large numbers, 394–395
memory group, 193
by multiples of ten, 473–475
multiples of 10's, 11's, and 12's, 194
multiples of 10's and 100's, 248–249
multiplication algorithm, 258
relation to division, 225–228
repeated addition, 258
sign, 150
of three-digit numbers, 732–733
of three-digit numbers by two-digit numbers, 705–706
of two-digit numbers, 253–255, 572–573, 680
of two two-digit numbers, 572–574, 581–583
using tens and ones, 258
word problems, 260–261, 475

Multiplication table
concept of, 151
using to divide, 227

N _____

Naming. *See* **Renaming; Classification**

Negative numbers, 130

Nets, 640, 642, 717

Nines, in multiplication, 162

Nonprime numbers. *See* **Composite numbers**

Notation. *See* **Expanded form**

Number lines. *See also* **Graphs**
decimals on, 611
fractions on, 610–611
rounding with, 266, 307
tenths and hundredths, 288–290
tick marks on, 128
two-step word problems, 617
using, 128–133

Number Sense (mental math), 7, 13, 17, 23, 27, 31, 37, 41, 46, 53, 63, 67, 73, 81, 89, 95, 101, 109, 115, 121, 135, 141, 149, 156, 161, 167, 173, 180, 187, 193, 205, 213, 219, 225, 233, 239, 247, 253, 281, 287, 293, 305, 311, 318, 323, 329, 343, 349, 376, 381, 387, 393, 409, 417, 423, 431, 436, 441, 447, 451, 457, 463, 473, 479, 487, 493, 499, 505, 513, 517, 525, 531, 541, 547, 555, 561, 567, 572, 577, 581, 586, 593, 603, 609, 615, 622, 627, 633, 639, 646, 653, 659, 669, 674, 681, 689, 695, 700, 705, 710, 721, 732, 737, 747

Number sentences. *See* **Equations**

Number systems, 174

Numbers. *See also* **Decimals; Mixed numbers; Whole numbers**
composite, 377
counting, 8
even, 9
large, reading and writing, 312–314
missing, 37–38, 43
negative, 130

odd, 9
positive, 130
prime, 376–377

Numerators. *See also* **Fractions**
defined, 115
reducing fractions, 701–702

O _____

Obtuse angles, 103–105

Obtuse triangles, 542

Octagonal prisms, 711

Octagons, 123, 331

Odd number sequences, 9

Online resources, Real-world investigations, 302, 408, 549

One
fractions equal to, 647
in multiplication, 157

Operations. *See* **Inverse operations; Order of Operations**

Order of Operations
concept of, 47–49, 241–243, 424–426
parentheses in, 167
writing expressions with, 241–243

Ordered pairs
coordinates of points, 470
defined, 468
graphing, 667

Ordering numbers
and comparing fractions, 382–383
decimal numbers, 277, 611
equivalent decimals, 295
fractions, 610–612
hundred thousands, 175
numbers in millions, 313, 324–326
by using place-value alignment, 20

Origin, 468

Ounce (oz), 532

Outcomes, 721–723

P _____

Parallel lines, 102

Parallelograms, as quadrilaterals, 594

Parentheses
Associative Property, 167–170
in coordinates, 468
in order of operations, 47
simplifying with, 424–426

Pentagonal prisms, 711

Pentagons, 123, 330–331

Percent (mental math), 343, 349, 393

Perfect squares, 203

Perimeter
activities about, 202, 206

INDEX

Two-step equations, 432

Two-step problems
    fraction of a group, 622–624
    word problems, 616–618

Twos, facts, 157

U

Unit multipliers. *See* **Conversion**

Units of measure. *See also* **Measurement**
    capacity, 487–489
    metric system, 283, 488, 533
    U.S. Customary System, 283, 487, 532

Unknown numbers, 48–49. *See also* **Missing numbers**

Unlikely events, 721–723

U.S. Customary System, 283, 487–488, 532. *See also* **Units of measure**

Use Logical Reasoning, 5

V

Vertex (vertices)
    of angles, 103
    of solids, 635

W

Weight *vs.* **mass,** 532–534

Whole numbers
    changing improper fractions to, 653–656
    factoring, 695–697
    fractions to nearest, 648
    names of, 18
    rounding, 306–308, 317, 648, 689–691
    through hundred millions, 689–691
    through hundred thousands, 174–176
    writing, 18–20, 174–176

Width, 122

Word problems
    about combining, 68–70
    about comparing, 96–98
    about division, 356–357, 452–454, 457–459, 481–482, 501, 527
    about fractions of a group, 494–495
    about multiplication, 260–261, 475
    about rates, 388–389, 400–401
    about separating, 90–93
    involving decimals, 307–308
    with remainders, 577–578
    two-step, 616–618
    writing, 59–62

Work Backwards, 5

Write a Number Sentence or Equation, 5

Writing about mathematics
    explain, 16, 22, 30, 34, 55, 57, 66, 72, 88, 106, 114, 119, 120, 125, 126, 139, 146, 159, 160, 179, 185, 186, 190, 191, 209, 221, 237, 243, 245, 250, 251, 257, 261, 262, 263, 290, 297, 303, 308, 315, 316, 328, 334, 335, 336, 337, 338, 353, 358, 360, 365, 373, 378, 385, 395, 402, 405, 407, 412, 418, 426, 427, 428, 429, 433, 440, 445, 456, 462, 466, 467, 470, 483, 484, 486, 492, 496, 509, 515, 522, 535, 544, 566, 569, 571, 583, 597, 598, 600, 607, 613, 614, 620, 625, 637, 643, 656, 658, 663, 668, 672, 676, 680, 691, 692, 697, 704, 713, 726, 727, 728, 734, 741
    exponents, 438
    expressions, 429–430
    large numbers, 312–313
    numbers through 999, 18–20
    numbers through hundred thousands, 174–176
    and problem-solving, 6
    and reading numbers in millions, 312–314
    word problems, 59–61

X

***x*-axis**
    coordinate on, 600
    defined, 468

×-symbol for multiplication. *See* **Symbols and signs**

Y

***y*-axis**
    coordinate on, 600
    defined, 468

Yard (yd), 283

Z

Zero
    dividing by, 232
    division answers, 500–501, 556–557
    in multiplication, 157, 248–249, 473–475, 568–569
    as placeholder, 573
    subtracting across, 82–84, 87–88